The Essentials of Composite Materials

The Essentials of Composite Materials: A Guide for Engineering and Beyond combines the theory of composite materials and their applications, with a focus on the main industries where they are used. Using the author's experience as a naval architect, boat builder, and composites designer, this book offers a guide to the selection of the most appropriate production processes, procedures, and materials for a particular project. It comprehensively covers polymer matrix composites, explaining what composite materials are, their components, and what they can be used for.

- Combines theoretical material with practical examples in a uniquely accessible way.
- Explores fabric structures, materials, resins, procedures, and manufacturing processes, including details that can only be discovered through hands-on work.
- Covers the more analytical side, explaining classical laminate plate theory, composite systems, strength, and failure criteria.
- Discusses applications in automotive, aerospace, civil, medical device, and naval industries.

This text serves as a practical tool for readers working in the composite fields as well as those looking to enter it.

Germán A. Pacheco is a naval architect, yacht designer, and composite technician with a vast experience in the marine and aerospace industry. He studied Naval Architecture at the National University of Quilmes in Argentina; he has worked in Australia and other countries for leading naval and aerospace companies. His experience includes the design and construction of vessels—from racing yachts to ferries, catamarans, cargo ships, and patrol boats—as well as the development of composite components for the marine, aerospace, and agricultural sectors. He has also contributed to the field through teaching everything he knows from his experience related to composite materials.

The Essentials of
Composite Materials
A Guide for Engineering and Beyond

Germán A. Pacheco

CRC Press
Taylor & Francis Group
Boca Raton London New York

CRC Press is an imprint of the
Taylor & Francis Group, an **Informa** business

Designed cover image: © 2026 Shutterstock

First edition published 2026
by CRC Press
2385 NW Executive Center Drive, Suite 320, Boca Raton FL 33431

and by CRC Press
4 Park Square, Milton Park, Abingdon, Oxon, OX14 4RN

CRC Press is an imprint of Taylor & Francis Group, LLC

© 2026 Germán A. Pacheco

ISBN: 978-1-032-93279-8 (hbk)
ISBN: 978-1-032-93286-6 (pbk)
ISBN: 978-1-003-56522-2 (ebk)

DOI: 10.1201/9781003565222

Typeset in Times
by Apex CoVantage, LLC

Contents

Preface

Many of our modern technologies and materials are the result of constantly seeking industry's needs. For example, aircraft engineers are frequently searching for structural materials with unique properties that cannot be met by conventional materials such as metal alloys, ceramics, or polymeric materials. These demands have pushed the development of material property combinations, which for the last decades have been extended by the creation of composite materials.

Composite materials have revolutionised the landscape of modern engineering, offering unparalleled versatility, strength, and adaptability. Their ability to combine the best properties of multiple materials into a singular, high-performance component has made them unique across several industries, from aerospace, marine, and automotive to construction and renewable energy. In this book, *The Essentials of Composite Materials: A Guide for Engineering and Beyond*, the readers will embark on a deep exploration of the principles, classifications, applications, and advanced methodologies that define the world of composite materials.

The book begins by establishing a solid foundation in the basics of composite materials, defining key concepts and exploring the roles of the matrix and reinforcement. This understanding is essential, as it promotes further discussions into the composite systems and their diverse applications, from their complex stress-strain relationships between the matrix and reinforcement to the interactions that settle their behaviour under load conditions.

Beyond the foundational concepts, *The Essentials of Composite Materials: A Guide for Engineering and Beyond* navigates into advanced topics such as composite structures, manufacturing techniques, and failure criteria. The manufacturing process is particularly critical, as it directly influences the quality and performance of the final composite product. From hand lay-up and filament winding to automated methods like resin transfer moulding, diverse techniques used to create composite materials are explored, emphasizing the importance of precision and innovation during these processes.

The book also addresses the theoretical frameworks, such as classical laminate theory and plate analysis. These methodologies are vital for understanding and predicting the mechanical performance of composite materials in real-world applications. By breaking down elaborate theories, this book simplifies complexity into clear and understandable explanations, ensuring that readers with basic background knowledge can comprehend and incorporate these essential concepts. Understanding failure criteria is equally crucial; the book addresses modes of failure, such as matrix cracking, fibre breakage, and delamination, and presents strategies to mitigate these risks through careful design and material selection.

Whether you are an experienced engineer, a student, or a professional seeking to deepen your knowledge, *The Essentials of Composite Materials: A Guide for Engineering and Beyond* serves as an indispensable resource. By bridging the gap between theory and practice, it provides readers with the tools needed to harness the full potential of composite materials.

1 What Exactly Is a Composite Material?

1.1 DEFINITION

To start from the very beginning, it is important to settle a correct and clear definition of composite materials. To explain it in simple words, a composite material is a combination of two or more constituent materials with different physical or chemical properties (different shapes and different compositions) that are essentially insoluble in each other. When these two are combined together, they provide a new material with different characteristics from the original properties.

Looking at it from the engineering aspect, when combined, they produce a material with properties that exceed the constituent materials, mechanically and physically. Looking at it closer, and speaking from the structural applications, it will generally have a rigid and resistant aspect, called REINFORCEMENT, and another one with less rigidity and resistance, called MATRIX. A stress-strain relation between these two materials can be plotted on a graph, including the resulting composite (see Figure 1.1).

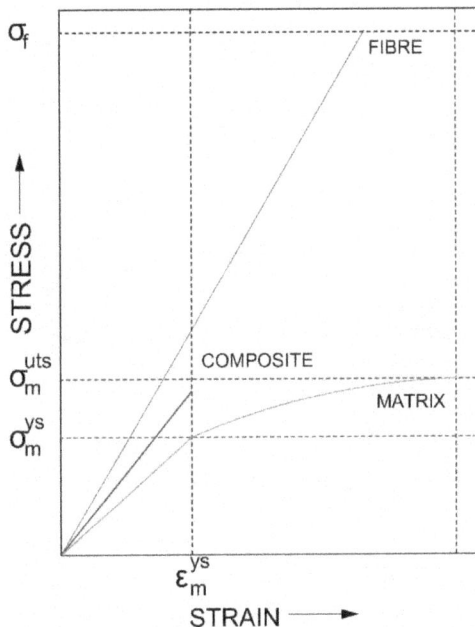

ELASTIC-BRITTLE FIBRE & ELASTIC-PLASTIC MATRIX

FIGURE 1.1 Stress-strain relationship of matrix, reinforcement, and the resulting composite.

DOI: 10.1201/9781003565222-1

When the search for the appropriate material begins, there are a number of factors to consider that will be of greater influence.

What do we normally look for in a material?:

- Design flexibility.
- Low weight.
- High impact strength.
- Low cost.
- Durability.
- Resistance to fatigue.
- Resistance to corrosion.
- Resistance to aggressive environments

1.2 COMPOSITE MATERIAL

People used to associate the words "composite materials" with carbon fibre, fibre-glass or even Kevlar, and, of course, resins like epoxy, polyester, and vinylester. These concepts are not completely wrong, but it is important to understand that composite materials themselves are not just that.

Most composite materials are conformed by two or more phases, as shown in Figure 1.2, a continuous matrix enclosing the dispersed phases that are classified according to their microstructure or geometry.

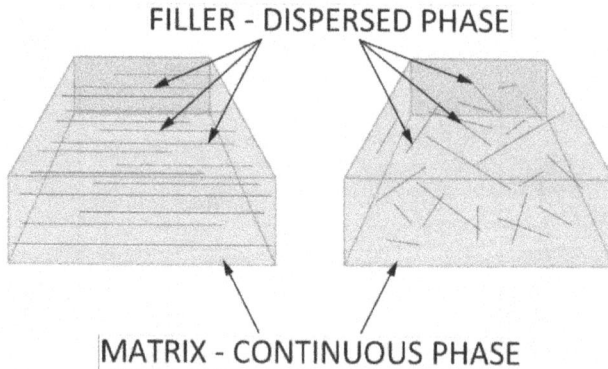

FIGURE 1.2 Dispersed phases contained in a continuous phase called matrix.

An example of a composite material known to all of us and daily present in nature is wood. Although there are very different types of wood with different mechanical properties of elasticity and deformation, such as particleboard, plywood, oriented strand lumber, and composite decking, see Figure 1.3.

As mentioned, composite materials can be formed by multiple combinations and a large variation of materials, which makes it really hard to put it all in a single box.

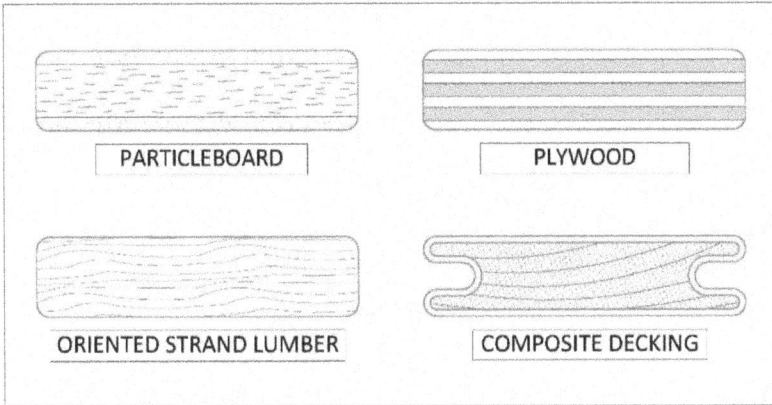

FIGURE 1.3 Types of wood with different properties of elasticity and deformation.

Due to this, it is important to know that composite materials can be classified based on two main types:

1) Referred to the Reinforcement form or geometry:
   - Fibre-Reinforced composites.
   - Laminar composites.
   - Particulate composites
2) Type of Matrix

1.2.1 FIBRE-REINFORCED COMPOSITES

Fibre-reinforced composites are the most important composites from a technological point of view. The main purpose of fibre-reinforced composites is to obtain materials with a high fatigue and rigidity resistance, at low and high temperatures, but at the same time getting a low density.

Putting these factors together, it is likely to say that it will achieve a better resistance-to-weight Ratio. This relation is possible due to the use of light materials, both in the matrix and in the fibres. Figure 1.4 shows a scaled representation of a random fibre-reinforced composite, where all the fibre reinforcements are placed or pointing in a single direction (unidirectional).

1.2.2 LAMINAR COMPOSITES

Laminar composites are products of conformed layers of materials held together by a matrix. They are made up of both composite and homogeneous materials, with the condition that they depend not only on the constituent materials but also on the design geometry of the structural elements.

FIBRE-REINFORCED COMPOSITE

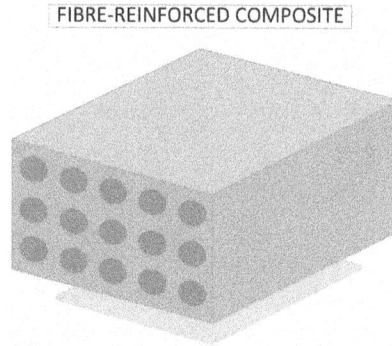

FIGURE 1.4   Schematic representation of a random fibre-reinforced composite.

The layers are stacked together one on top of the other, in which the orientation of the high-strength direction varies within each layer.

This type of reinforcement form could also be known as a sandwich structure. Figure 1.5, for example, represents two layers of material bonded together and held by a matrix.

LAMINAR COMPOSITE

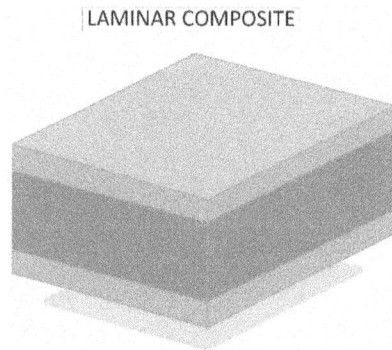

FIGURE 1.5   Schematic representation of a laminar composite.

Laminar composites are classified as follows:

- Laminar compounds.
- Sandwich structures.
- Non-laminar structures

1.2.3 PARTICULATE COMPOSITES

Particulate composites, as the name itself suggests, are made of particles distributed in a matrix body. Basically, these materials are made by combining a matrix with the mentioned particles known as fillers, which could be flakes or in powder form.

The particles are added randomly to the matrix (Figure 1.6), which makes these composites isotropic. In essence, the matrix transfers part of the present stress to these particles, which carry part of the load. Concrete and wood particle boards are examples of this category.

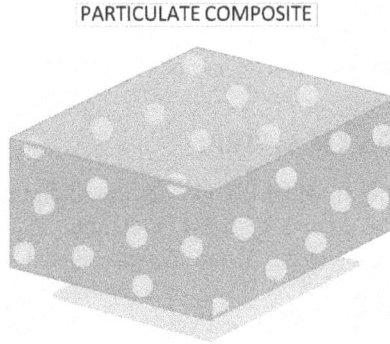

PARTICULATE COMPOSITE

FIGURE 1.6   Schematic representation of particulate composite.

1.3   WHAT IS A MATRIX?

The matrix is a continuous material in which the reinforcement is "contained". Metallic, ceramic, or organic resins can fulfil this role.

The main purposes of the matrix are as follows:

- Contain the reinforcement.
- Define the physical and chemical properties.
- Transfer the loads to the reinforcement

Furthermore, it will allow for setting some characteristics of the composite material, such as the way it conforms and its surface finish.

The properties of the matrix will depend on the ability of the composite material to be conformed to complex geometries. By subjecting or exposing the composite material to different types of stress and mechanical loads, the matrix will play different roles.

1) Under compressive loads: it is the matrix that supports it, since it is the continuous material.
2) In traction: the matrix transfers the load applied to each of the fibres or particles, so that these are the ones that support the load. To achieve this, it is necessary for them to have an excellent bond between the matrix and the reinforcement.

In addition, generally, it is the matrix that provides impact resistance and helps to stop crack propagation.

1.3.1 Matrix Properties

- Supports the fibres, keeping them in their correct position.
- Transfers the load to the strong fibres.
- Protects the composite from damage during manufacture and after long use.
- Prevents crack propagation in the fibres throughout the length of the composite.
- The matrix is generally responsible for the primary control of the electrical properties, chemical behaviour, and the use of the composite at high temperatures.

1.4 CLASSIFICATION OF MATRIX

There are three main types in which composites based on the type of matrix can be classified: polymer matrix composites (PMC), ceramic matrix composites (CMC), and metal matrix composites (MMC), and they are shown below in Figure 1.7 as a tree diagram representation:

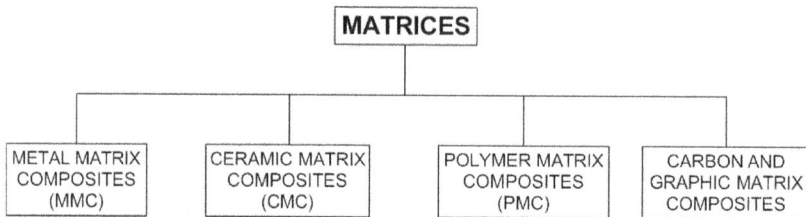

FIGURE 1.7   Classification scheme for the various matrix types.

1.4.1 Metal Matrix Composites (MMC)

Metal matrix composites offer high strength and resistance, fracture toughness and stiffness, high-temperature applications, and high thermal conductivity. Compared to polymer composites, they can withstand elevated temperatures while experiencing a corrosive environment.

Most metals and alloys can be used as a matrix, and they require reinforcement materials that need to be stable over a range of temperatures. These reinforcement materials also need to be non-reactive.

Most metals and alloys work well as a matrix. However, the choices for low-temperature applications are limited. Only light metals are responsive, having the advantage of providing low density like titanium, aluminium, and magnesium, which are particularly useful for aircraft applications.

If metallic matrix materials have to offer high strength, then they will require the application of high-modulus reinforcements. As an extra detail, the strength-to-weight ratios of resulting composites can be higher than most alloys.

1.4.2 Ceramic Matrix Composites (CMC)

Ceramics can be described as solid materials with very strong ionic bonding in general. High melting points, good corrosion resistance, stability at elevated temperatures, and high compressive strength are the main characteristics of these materials; however, they are fragile materials.

Ceramic-based matrix materials are favourite for applications requiring a structural material that does not give way at high temperatures.

1.4.3 POLYMER MATRIX COMPOSITES (PMC)

The next chapter will delve deeply into polymer matrix composites, but as a first definition, PMCs are the most commonly used composites. They are materials composed of synthetic reinforcing fibres conformed in a polymeric matrix or plastic resin. They present high strength resistance, and they are generally easy to process. Also, they have the significant advantage of lightweight and desirable mechanical properties.

BIBLIOGRAPHY

'Analysis and Performance of Fiber Composites' (Third edition). Bhagwan D. Agarwal, Lawrence J. Broutman and K. Chandrashekhara. Wiley India Pvt. Ltd., 2015.

'Composite Materials'. S. C. Sharma. Alpha Science International, Ltd, 2000.

'Composite Materials: Design and Applications'. Daniel Gay, Suong V. Hoa and Stephen W. Tsai. CRC Press LLC, 2003.

'Composite Materials Handbook. Volume I – Polymer Matrix Composites Guidelines for Characterization of Structural Materials – MIL-HDBK-17'. McGraw Hill/Departments and Agencies of the Department of Defence, 2002.

'Engineering Mechanics of Composite Materials' (Second edition). Isaac M. Daniel and Ori Ishai. New York: Oxford University Press, 2006.

'Fiber Composite Analysis and Design: Composite Materials and Laminates. Volume I'. Z. Hashin, B. W. Rosen, E. A. Humphreys, C. Newton and S. Chaterjee. Washington, DC: U.S. Department of Transportation: Federal Aviation Administration Office of Aviation Research, 1997.

'Handbook: An Engineering Compendium on the Manufacture and Repair of Fiber Reinforced Composites'. R. L. Ramkumar, N. M. Bhatia, J. D. Labor and J. S. Wilkes. NJ, USA: Department of Transportation FAA Technical Center: Atlantic City International Airport, 1987.

'Handbook of Composites' (Second edition). S. T. Peters. Mountain View, CA, USA: Process Research, 1997.

'Introduction to Composite Materials Design'. Ever J. Barbero. USA: Department of Mechanical & Aerospace Engineering – West Virginia University/Taylor & Francis, 1998.

'Mechanics of Composite Materials' (Second edition). Robert M. Jones. Taylor & Francis, 1999.

'Shigley's Mechanical Engineering Design' (Tenth edition). Richard G. Budynas and J. Keith Nisbett. McGraw Hill Education, 2015.

'Structural Composite Materials'. F. C. Campbell. ASM International, 2010.

'The Theory of Composites'. Graeme W. Milton. Cambridge: University of Utha Press, 2002.

2 Polymer Matrix Composites

Polymers make ideal materials as they can be processed easily, possess high strength and resistance, are lightweight, and have desirable mechanical properties. These are some of the main reasons why they are extensively used in aeronautical applications, for example.

They have an important influence on technology, mainly when they are reinforced with fibres, such as

- Carbon fibre-reinforced plastic (CFRP)
- Glass fibre-reinforced plastic (GFRP)
- Aramid fibre-reinforced plastic (AFRP)

2.1 ADVANTAGES AND DISADVANTAGES OF POLYMER MATRIX COMPOSITES

2.1.1 ADVANTAGES

Lightweight: Composites are lightweight compared to most metals. This property makes them ideal for the construction of yacht hulls, aircraft fuselages, interiors of many vehicles, bicycle frames, military purposes, and much more. It is a huge advantage where less weight means better efficiency in the final product.

Strength-to-Weight Ratio: Strength-to-weight ratio, as a definition, refers to the material's strength in relation to how much it weighs. Some materials are very strong and heavy, such as steel. Composite materials can be designed to be both strong and light.

This property is why composites are used to build airplanes, for example, where the need is to produce a very high strength material at the lowest weight possible.

Corrosion Resistance: Composites can resist damage from the weather, UV exposure, and harsh chemical attacks. When they are exposed outdoors to these extreme conditions, they can withstand severe weather and wide changes in temperature.

Design Flexibility: Composites can be moulded into complicated shapes more easily than most other materials. This gives designers the freedom to create almost any shape or form as desired.

Part Consolidation: A single piece made of composite materials can replace an entire assembly of metal parts. This is a great advantage since the number of parts in a machine or structure can be reduced, saving time and maintenance over the lifetime.

DOI: 10.1201/9781003565222-2

Dimensional Stability: Composites can retain their shape and size in different situations within a reasonable range. This means that their dimensions will not alter when they are hot or cold, wet or dry – no thermal expansion.

Durable: Structures made of composites have a long life and need less maintenance compared to other materials used in the industry, like metals. Highly resistant to fatigue, chemical corrosion, rust, and scratches are some of the main characteristics in terms of durability.

Resistance to seawater, for example, makes it best to be used in boat building, shipping, and marine industries.

2.1.2 DISADVANTAGES

Delamination: Since composites are often built from a stack of different plies of layers into a laminate structure, it is a reality that they can "delaminate" between layers if they find a weak point.

Decomposition: Composites are likely to decompose at high temperatures; this factor greatly influences and limits the service temperature.

High Cost: They are relatively new materials, and as such, they can have a high cost. Of course, it will depend on the specific materials chosen, the manufacturing process, etc. But the point mainly focuses on the necessity of having the right qualified personnel, which is not easy to find.

Complex Fabrication: The fabrication process is usually labour intensive and complex, which, as mentioned in the previous point, will further increase costs.

Damage Inspection: One of the most common types of failures in composites is "delamination", followed by "cracks". These types of failures are mostly internal and require complicated inspection techniques for detection.

Composite to Metal Joining: Metals expand and contract more with variations in temperature compared to composites. This may cause an imbalance when trying to couple both materials and may lead to failure.

2.2 CLASSIFICATION OF POLYMER MATRIX COMPOSITES

As mentioned, polymer matrix composites (PMC) belong to a family of matrices classification, but to continue, PMCs can also be classified into two other subcategories.

As shown in Figure 2.1, polymer matrix composites can be classified into thermosets and thermoplastics.

Within the thermosets family, we can find polyester, epoxy, vinylester, and polyamides.

While in thermoplastics we can find polyethylene (including HDPE and LLDPE/LDPE), polypropylene (PP), polyvinyl chloride (PVC), and polyethylene terephthalate (PET).

```
            ┌─────────────────────┐
            │   POLYMER MATRIX    │
            │    COMPOSITES       │
            │      (PMC)          │
            └─────────────────────┘
                      │
          ┌───────────┴───────────┐
          │                       │
 ┌────────────────┐      ┌──────────────────┐
 │  THERMOSETS    │      │  THERMOPLASTICS  │
 └────────────────┘      └──────────────────┘
```

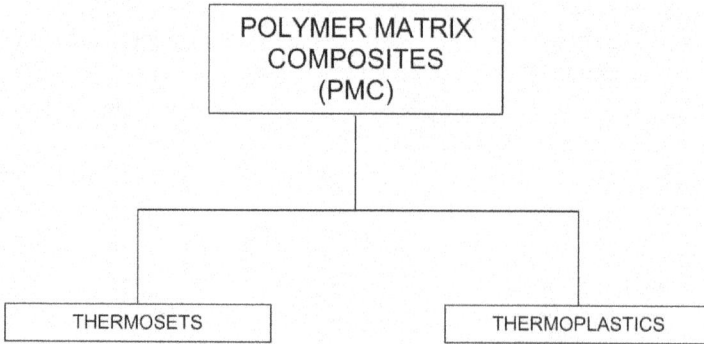

FIGURE 2.1 Polymer matrix composites classification scheme.

2.2.1 THERMOSET POLYMER MATRIX

Thermosets are the most popular of the fibre composite matrices used in aerospace components, automobile parts, defence systems, and more.

One of the critical points to have in mind about thermosets is that these materials are hard to prevail at elevated temperatures. They are considered a fragile material, and due to their non-melting condition, they cannot be reversed to regain their properties as other materials do. They also need controlled temperature storage.

Thermosets polymerize and link together during the cure cycle; this means that they are getting their final properties during the hardening stage. This is managed by the catalyst (hardener) and the application of heat.

Thermosets have qualities such as a well-bonded three-dimensional molecular structure after curing, slow creep resistance, easy processing, and desirable mechanical properties.

Another characteristic or advantage is that they decompose instead of melting at high temperatures.

The most popular thermosets used in the industry are as follows:

- **Polyester Resins:** These are quite easy to access or find, are cheap in comparison to other materials, and can be used in a wide range of fields or industries.

  Liquid polyesters can be stored at room temperature for months, sometimes for years, and the mere addition of a catalyst (hardener) can cure the matrix material within a short time. They are mostly used in automobile and structural applications, where the structure is not compromised.

  The cure of polyester is usually rigid or flexible, as the case may be, and tends to be transparent. Polyesters withstand variations in the environment and are stable against chemical attacks. Depending on the formulation of the resin, they can be used at temperatures up to about 75°C or higher.

Other advantages of polyesters include easy compatibility with a few glass fibres. On the other hand, polyester resins produce toxic styrene emissions.

- **Epoxy Resins**: When compared with polyester resins, they are widely used in filament-wound composites and are suitable for moulding prepregs. They have high mechanical and thermal properties and are reasonably stable against chemical attacks.

  In terms of bonding and final molecular structural composition, epoxy resins are excellent adherents, exhibiting slow shrinkage during curing and no emission of volatile gases.

  These advantages, however, make the use of epoxies rather expensive.

- **Vinylester Resins:** They are produced by the reaction (esterification) between an epoxy resin and an unsaturated monocarboxylic acid. Essentially, they have a base of polyester strengthened with epoxy molecules in the backbone of the molecular chain. Vinylester also uses peroxide (e.g., MEKP) for hardening.

  Before the addition of styrene, vinylesters are midway between polyester and epoxy in terms of viscosity and mechanical properties.

  Due to a few open sites in the molecular chain, the resistance to water penetration (hydrolysis) is higher.

  Vinylesters shrink less during curing, which means that the "pre-release" of a laminate from a mould is less significant. Vinylesters are more tolerant to stretching than polyesters, which makes them able to absorb impact without damage. They are also less likely to show stress cracking.

  The cross-bonding of vinylesters is superior to that of polyesters; the bond to core materials is much more effective than with polyesters, and delamination is less of an issue. As last, they are less sensitive to ambient conditions (temperature and humidity) than polyesters as well.

To put it in simple words and explain the difference between the three of them, polyesters are more user-friendly in terms of chemical preparation. It is important to take care when using the correct components' mixture, but it has a wide tolerance for that variation. We can control the working time based on the different percentages of component mixture, the environment, or even by applying external temperature influences. On the other hand, quality is very poor compared with vinylester and far from epoxy.

Epoxy is the best in terms of quality and mechanical properties, but it also makes it quite expensive, considering materials, processes, and the qualified labour needed to work with it.

2.2.2 THERMOPLASTIC POLYMER MATRIX

They are fully polymerized polymers. They can be physically altered with the application of heat.

Thermoplastics have one- or two-dimensional molecular structures, and they tend to prevail at elevated temperatures, showing an exaggerated melting point. Another advantage is that the process of softening at elevated temperatures can be reversed to regain their properties during cooling, facilitating the application of conventional compression techniques to mould the composites.

They are suitable for injection or thermoforming processes and present high toughness and ductility. However, in terms of cost, they can be little high.

Processes are not easy to control; high viscoelasticity could be a problem, and they present a short life when exposed to fatigue.

2.3 MOLECULAR CONFIGURATION OF THE POLYMER MATRIX COMPOSITES

To describe their composition, polymer materials have molecular structures with long chains made of small repeating units joined together end to end. As a side note, the word "polymer" comes from the Greek, where poly means (many) and "mer" or "meros" means (part).

Many engineering polymer materials are organic composites based on carbon, hydrogen, and other non-metals. Polymers are in general low in density and flexible compared to other materials.

Figure 2.2 shows, for example, a polyethylene molecular structure (known as PE), where it can be appreciated how the units repeat end to end.

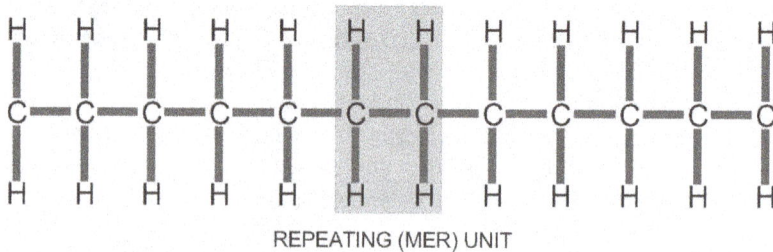

REPEATING (MER) UNIT

FIGURE 2.2   Polyethylene molecular structure.

As more repeating molecular units are added, the polyethylene molecular chain length grows and extends, and the same will happen with the molecular weight (more precisely, molecular mass or molar).

2.3.1   MECHANICAL PROPERTIES OF THE POLYMER

The mechanical properties that the polymer will show depend directly on two factors.

1) Length of the molecule
2) Shape of the molecule

2.3.1.1   Molecular Length

The joining of the repeating molecular units creates a chain; polymers with very long chains present an extremely large molecular weight. Not all the polymer chains grow to the same length; as the number of repeating units increases, both the physical polymer chain length and the referred molecular weight of the polymer will increase as well.

2.3.1.2   Molecular Structure

Polymers are giant molecules or macromolecules with covalently bonded carbon atoms as the backbone of the chain. Small-chain, or low molecular-weight organic molecules that we can call just (monomers) are joined together via the process of polymerization, which converts monomers to polymers.

2.3.2   Molecular Configuration

There are several molecular configurations in polymers. The basic configurations are given below:

2.3.2.1   Linear Polymer

Linear polymers are basically long chain-like structures, as shown in Figure 2.3, where repeating units are joined together end to end in a "line".

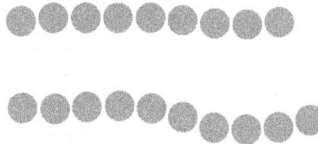

LINEAR POLYMERS

FIGURE 2.3   Representation of a linear polymer chain structure.

2.3.2.2   Branched Polymers

Branched polymers feature short side-branched chains that are connected to the main (longer) polymer chains, as shown in Figure 2.4. The packing efficiency of the polymer with side branching results in low polymer density.

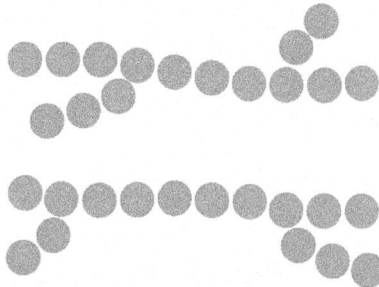

BRANCHED POLYMERS

FIGURE 2.4   Representation of a branched polymer chain structure.

2.3.2.3 Cross-linked Polymers

The linear polymer chains of cross-linked polymers are joined together via covalent bonding of smaller molecules acting as bridges between them. Figure 2.5 shows bridging connections between linear polymer chains. This makes the polymer strong and rigid. The cross-linking process often takes place through a non-reversible chemical reaction process (curing).

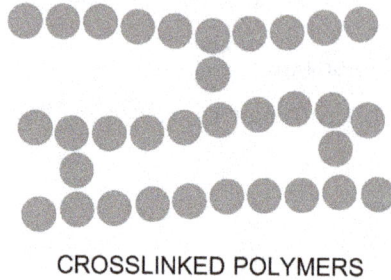

CROSSLINKED POLYMERS

FIGURE 2.5   Representation of a cross-linked polymer chain structure.

2.3.2.4 Networked Polymers

Networked polymers, as shown in Figure 2.6, are characterised by their repeating units with three active covalent bonds, creating an interconnected three-dimensional network configuration. Networked polymers include highly cross-linked polymers. For example, highly cross-linked epoxies often present a highly cross-linked three-dimensional epoxy network configuration.

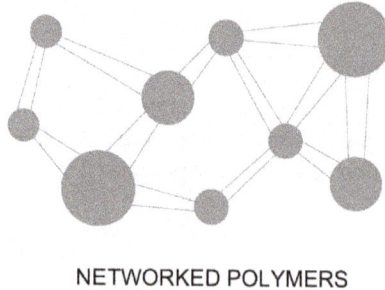

NETWORKED POLYMERS

FIGURE 2.6   Representation of a networker polymer chain structure.

2.3.3 Thermoset Polymer Molecular Structure

Thermoset or thermosetting polymers are the result of the molecular cross-link network of covalent bonds that are created between primary polymer chains. This takes place during polymerization and cross-linking during a curing cycle with the help of a hardening agent. Heating is another factor to consider as a promoter.

During curing, their physical composition changes from a viscous fluid to a rub-bery gel (viscoelastic material) and finally to a glassy solid. If heated after curing, they initially become soft and rubbery at high temperatures, and if further heated, they will not melt but will decompose (burn).

In terms of molecular structure, the polymer chains do not have the physical abil-ity to re-organise into packed and folded molecular chain structures; they effectively become irreversibly hardened after curing and cannot be reprocessed, making them single-use and not recyclable.

Popular examples of thermoset polymers include epoxy and polyester.

2.3.4 THERMOPLASTIC POLYMER MOLECULAR STRUCTURE

They are generally linear or branched in structure. When heated, thermoplastics soften and melt, flowing as a viscous liquid. This process can be repeated, making them potentially recyclable.

BIBLIOGRAPHY

'Analysis and Performance of Fiber Composites' (Third edition). Bhagwan D. Agarwal, Law-rence J. Broutman and K. Chandrashekhara. Wiley India Pvt. Ltd., 2015.

'Composite Materials'. S. C. Sharma. Alpha Science International, Ltd, 2000.

'Composite Materials: Design and Applications'. Daniel Gay, Suong V. Hoa and Stephen W. Tsai. CRC Press LLC, 2003.

'Composite Materials Handbook. Volume I – Polymer Matrix Composites Guidelines for Characterization of Structural Materials – MIL-HDBK-17'. McGraw Hill/Departments and Agencies of the Department of Defence, 2002.

'Engineering Mechanics of Composite Materials' (Second edition). Isaac M. Daniel and Ori Ishai. New York: Oxford University Press, 2006.

'Fiber Composite Analysis and Design: Composite Materials and Laminates. Volume I'. Z. Hashin, B. W. Rosen, E. A. Humphreys, C. Newton and S. Chaterjee. Washington, DC: U.S. Department of Transportation: Federal Aviation Administration Office of Aviation Research, 1997.

'Handbook: An Engineering Compendium on the Manufacture and Repair of Fiber Reinforced Composites'. R. L. Ramkumar, N. M. Bhatia, J. D. Labor and J. S. Wilkes. NJ, USA: Department of Transportation FAA Technical Center: Atlantic City International Air-port, 1987.

'Marine Composites' (Second edition). Eric Green Associates. MD: Eric Green Associates Inc., 1999.

'Mechanics of Composite Materials' (Second edition). Robert M. Jones. Taylor & Francis, 1999.

'The Theory of Composites'. Graeme W. Milton. Cambridge: University of Utha Press, 2002.

3 Fillers or Type of Reinforcements

3.1 REINFORCEMENTS AND FILLERS

There is a little difference between the reinforcement and the filler. In terms of mechanical properties, the reinforcement can improve the tensile and flexural strength when applied, while the fillers do not. The main condition is that the reinforcement must create a strong adhesive bond with the resin to ensure the correct result.

Composite fillers and reinforcements are mostly used to modify or improve the physical and mechanical properties of plastics. Fillers and reinforcements may also be used to lower the material costs by reducing the volume of resin (matrix) required.

3.2 FILLERS

The fillers are mostly used to be combined with resin to bulk it, but they can also improve slightly the compressive strength of the material applied, reducing potential shrinkage during the cure cycle.

Fillers are also in charge of modifying properties such as thermal conductivity, electrical resistivity, and friction, and they can also reduce the material cost.

The use of fillers can alter the polymer properties in the following ways:

- Lower shrinkage.
- Increase in hardness or stiffness.
- Increase in modulus of elasticity.
- Increase in density.
- Increase thermal resistance.
- Increase abrasion resistance.
- They have a favourable coefficient of thermal expansion.
- Impact colour or opacity to the composite which they fill.

3.2.1 ADVANTAGES AND DISADVANTAGES OF FILLERS

Methods of fabrication are very limited, and the curing of a few resins is very constrained, shortening the lifespan of some resins and weakening a few composites. In these cases,

- Fillers could be the main ingredient or an additional one in composite materials. The filler particles could have a form of irregular structures or have a precise shape like polyhedrons, short fibres, or spheres.
- An inert addition, fillers can modify almost any basic resin characteristics in all directions required. The final composite properties can be affected by

DOI: 10.1201/9781003565222-3

the shape, surface treatment, size of the particles in the filler material, and the size distribution.

- Fillers produced from powders are also considered particulate composites.
- In the honeycomb structure, for example, sheet materials in hexagonal shapes are impregnated with resin mixed with filler to ensure a high-performance bonding. Foam materials are also used as a core in sandwich composites.

Figure 3.1 reflects a joining procedure for a sandwich structure, having a core material (in this case, honeycomb), followed by the adhesive on each side of the core (this could be filler mixed with resin) and both face sheets on the top and bottom of the sandwich. All are bonded together as a result of the adhesive.

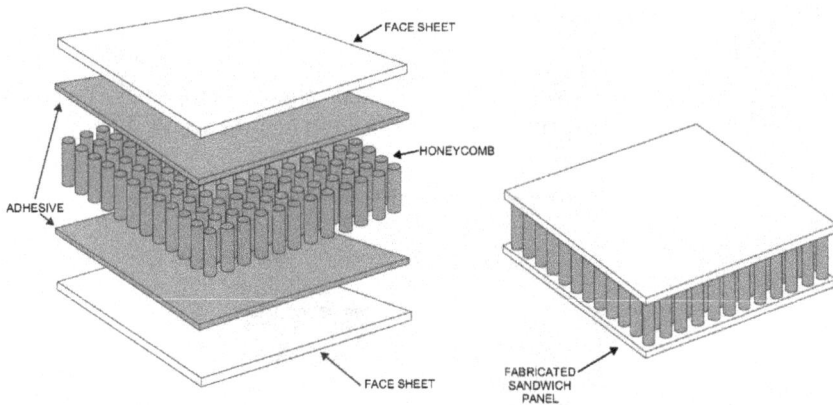

FIGURE 3.1   Sandwich panel structure bonding procedure.

3.2.2 CLASSIFICATION OF FILLERS

Fillers can be classified as

- Particles
- Microspheres, which are also subdivided into solid and hollow

Figure 3.2 shows a representation scheme of filler classification.

3.2.2.1 Particle Fillers

Particulate fillers are often added to polymers to increase stiffness.

Desirable properties for filler particles are reasonable stiffness (though this is much less crucial than with fibres), good interfacial bonding with the polymer, chemical and thermal stability, and low cost.

3.2.2.2 Microspheres

Microspheres are the most useful fillers due to their specific gravity and stable particle size. They also have considerable strength and controlled density, which allows

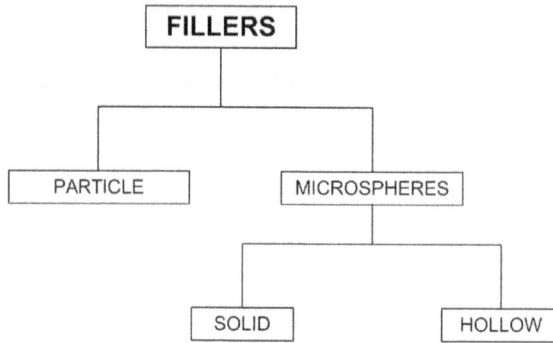

FIGURE 3.2   Filler's classification scheme.

them to modify products without compromising physical properties. Figure 3.3 shows the microspheres magnified under the electron microscope view.

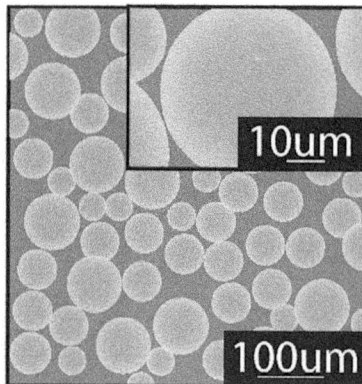

FIGURE 3.3   Microspheres representation under the electron microscope at 250 x magnification.

3.2.2.3   Solid Glass Microspheres

Solid glass microspheres are most suitable for plastics, including an excellent bond between the sphere's surface and resin. This type of filler can increase the bonding strength and reduce absorption of moisture (reduce attraction between particles). A great advantage of solid microspheres is their relatively low density.

3.2.2.4   Hollow Microspheres

They have controlled specific gravity, are larger than solid glass spheres, and are mostly used in polymers due to their wide particle size. This type of filler is less sensitive to moisture than the others.

To enclose all the specific fillers mentioned before, the particle-filled mix with resin is mostly used, for example, in addition to a better and stronger bond between composites. Let us say a composite bulkhead needs to be laminated, and it is made by a foam core enclosed by two laminate skins on each side (sandwich), something similar to the example shown in Figure 3.1. This bonding can definitely be improved by using a mix of particle-filled and epoxy resin. The same procedure could be used in the creation of a composite rudder, hull, or deck construction.

Speaking about the microsphere, this is mostly used as an addition or help on specific surface finishes. They are relatively easy to manipulate and easier in terms of shaping compared with the particles that are stiffer.

3.3 REINFORCEMENTS

Fibre reinforcements significantly affect the properties of the composites to which they are added. Reinforcements are particulates, fibres, or fabrics used to strengthen or toughen plastics, metals, or ceramics.

There are different types of fibre reinforcements, and they can be classified based on their geometry, as shown in Figure 3.4.

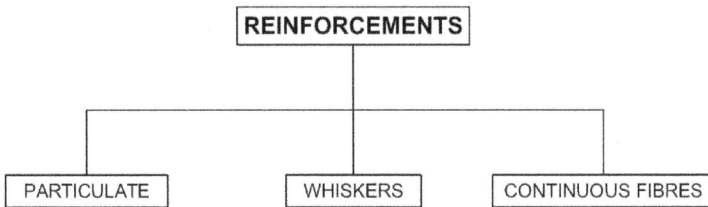

FIGURE 3.4 Reinforcement classification scheme

On the other hand, Figure 3.5 shows the most common reinforcement forms based on the type of reinforcement applied to composites.

3.3.1 PARTICULATE REINFORCEMENTS

Made by particles with different shapes and sizes, aleatory dispersed in the matrix. Figure 3.6 shows a magnified view of particulate composite.

Due to this aleatory distribution, they can be considered a quasi-isotropic material

Examples of particulate composites are concrete and aluminium particles in polyurethane.

3.3.2 DISCONTINUOUS FIBRES OR WHISKERS

These types of reinforcements are made by short fibres or whiskers. The fibres are longer compared with their diameter and their orientation could be aleatory, as shown in Figure 3.7, or unidirectional, as shown in Figure 3.8. They are generally used in applications exposed to low mechanical stress.

Due to this aleatory distribution, they can be considered a quasi-isotropic material.

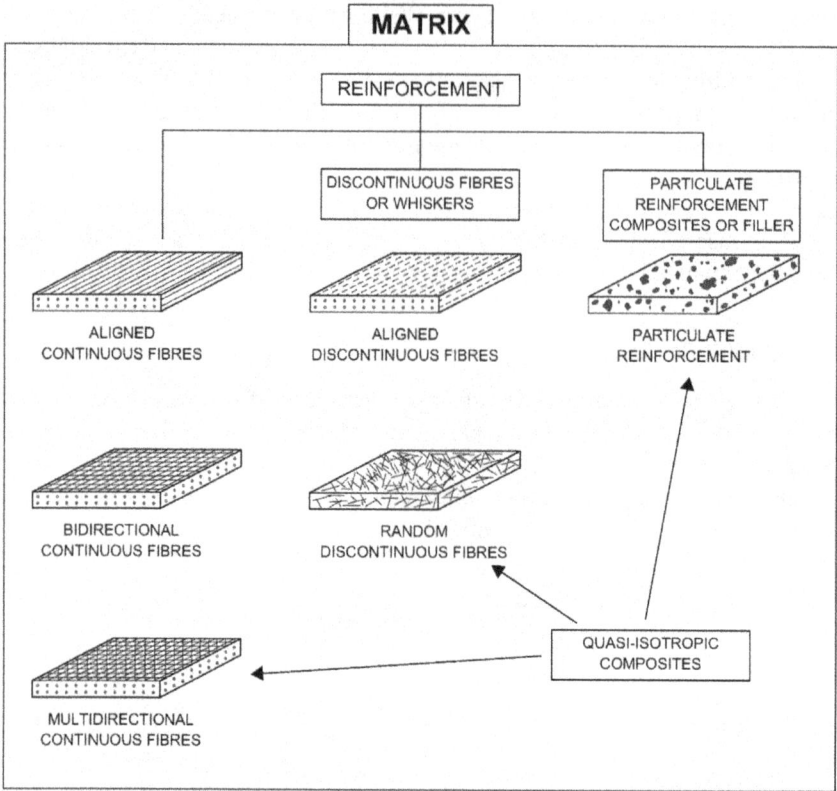

FIGURE 3.5   Classification of composites based on the type of reinforcement.

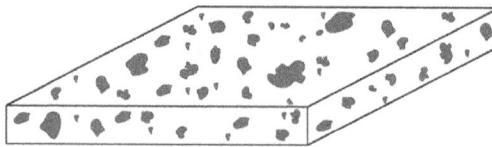

FIGURE 3.6   Representation of particulate reinforcement.

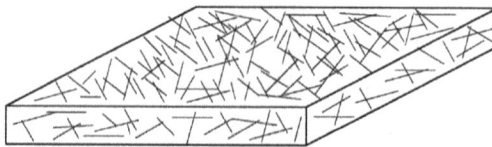

FIGURE 3.7   Random discontinuous fibres.

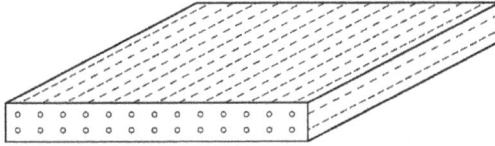

FIGURE 3.8 Aligned discontinuous fibres.

3.3.3 CONTINUOUS FIBRES

Reinforcement made by long and continuous fibres can be found both dry and pre-impregnated with resin. These continuous fibres could have fibre orientation as unidirectional (see Figure 3.9), bidirectional (see Figure 3.10), or multidirectional (see Figure 3.11). They are generally used where high rigidity and/or resistance are required.

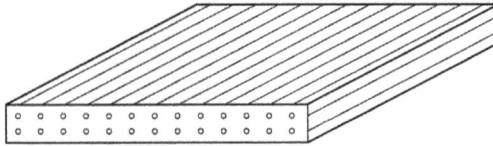

FIGURE 3.9 Aligned continuous fibre reinforcement.

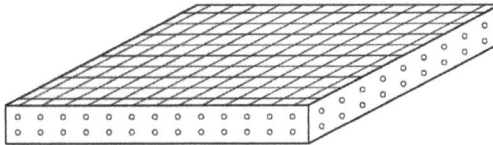

FIGURE 3.10 Bidirectional continuous fibre reinforcement.

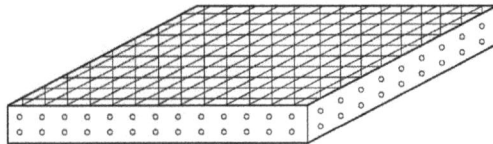

FIGURE 3.11 Multidirectional continuous fibre reinforcement.

They can be found as a roving appearance (winding), tape (only the fibres in one direction), or woven (fibres intertwined in more than one direction).

BIBLIOGRAPHY

'Analysis and Performance of Fiber Composites' (Third edition). Bhagwan D. Agarwal, Lawrence J. Broutman and K. Chandrashekhara. Wiley India Pvt. Ltd., 2015.

'Composite Materials: Design and Applications'. Daniel Gay, Suong V. Hoa and Stephen W. Tsai. CRC Press LLC, 2003.

'Engineering Mechanics of Composite Materials' (Second edition). Isaac M. Daniel and Ori Ishai. New York: Oxford University Press, 2006.

'Fiber Composite Analysis and Design: Composite Materials and Laminates. Volume I'. Z. Hashin, B. W. Rosen, E. A. Humphreys, C. Newton and S. Chaterjee. Washington, DC: U.S. Department of Transportation: Federal Aviation Administration Office of Aviation Research, 1997.

'Handbook of Composites' (Second edition). S. T. Peters. Mountain View, CA, USA: Process Research, 1997.

'Handbook of Fillers for Plastics'. Harry S. Katz and John V. Milewski, Eds. New York: Van Nostrand Reinhold, 1987.

'Mechanics of Composite Materials' (Second edition). Robert M. Jones. Taylor & Francis, 1999.

'Structural Composite Materials'. F. C. Campbell. ASM International, 2010.

4 Role of Fibres and the Matrix

The main role of fibres in composite materials is to enhance strength and stiffness. Usually made from high-strength materials, these fibres are embedded in a polymer matrix.

In a composite, the fibres handle the majority of the load due to their superior strength and stiffness, while the polymer matrix binds the fibres and helps to distribute the load among them.

4.1 ROLE OF FIBRES

Fibres play an important role in the creation of a composite material, as they can carry in-plane loads, providing stiffness and strength, and they also reduce the coefficient of thermal expansion (CTE).

Fibres are in charge of modifying properties such as thermal conductivity, electrical resistivity, and friction. They will also help to reduce the material cost.

4.1.1 FIBRE AXIAL STIFFNESS AND STRENGTH

When discussing stiffness properties, also referred to as the elastic properties, this will include the modulus of elasticity (E), the shear modulus (G), and Poisson's ratio (v).

Stiffness properties are independent of the material orientation, and only one value exists for each of these three stiffness properties. In contrast, the stiffness properties of unidirectional fibre-reinforced composites are highly dependent on the fibre orientation in relation to the applied force. Figure 4.1 shows a representation of these independent properties in each direction: 1, 2, and 3.

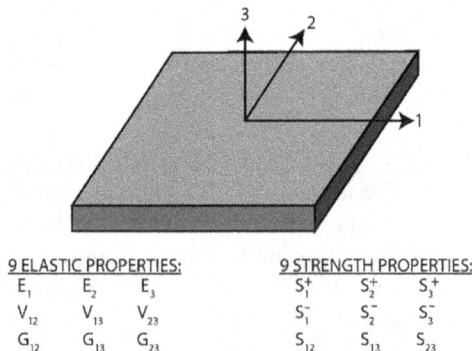

9 ELASTIC PROPERTIES:

E_1	E_2	E_3
v_{12}	v_{13}	v_{23}
G_{12}	G_{13}	G_{23}

9 STRENGTH PROPERTIES:

S_1^+	S_2^+	S_3^+
S_1^-	S_2^-	S_3^-
S_{12}	S_{13}	S_{23}

FIGURE 4.1 Independent elastic and strength properties of a unidirectional composite.

DOI: 10.1201/9781003565222-4

The values of each stiffness property will depend on the three perpendicular material orientations that are relative to the load applied.

However, when multiple layers at $0°$ are cured together to make a multilayer unidirectional composite, the random distribution of fibres between the layers produces multiple transverse stiffness values in the plane perpendicular to the fibre direction, as shown in Figure 4.2, assuming transverse isotropy.

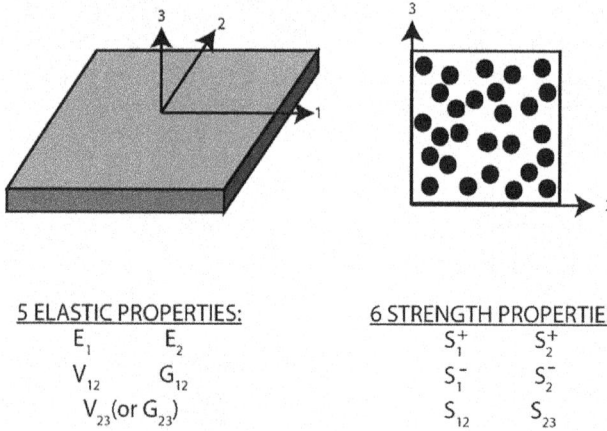

5 ELASTIC PROPERTIES:

E_1 E_2

V_{12} G_{12}

V_{23} (or G_{23})

6 STRENGTH PROPERTIES:

S_1^+ S_2^+

S_1^- S_2^-

S_{12} S_{23}

FIGURE 4.2   Multilayer of a unidirectional composite.

In conclusion, the material will have the same stiffness, despite if it is pulled in a horizontal or vertical direction.

4.2   COEFFICIENT OF THERMAL EXPANSION (CTE)

To define the coefficient of thermal expansion in a simple way, CTE measures how much the length of a material changes when it is exposed to high temperatures or cooled over a specific temperature range. It is typically expressed as a coefficient per unit temperature change at a certain temperature. This property is crucial for materials, particularly in composite structures that experience temperature fluctuations.

Fibre-resin composite materials are specially selected for aerospace structures due to their near-zero CTE. The dependence of CTE on fibre volume is crucial, while the longitudinal CTE is virtually unaffected by any changes in the laminate fibre content.

Longitudinal and transverse CTEs of continuous fibre-reinforced composites can be examined. Figure 4.3 shows the variation of the coefficient of thermal expansion of a unidirectional glass fibre laminate (UD).

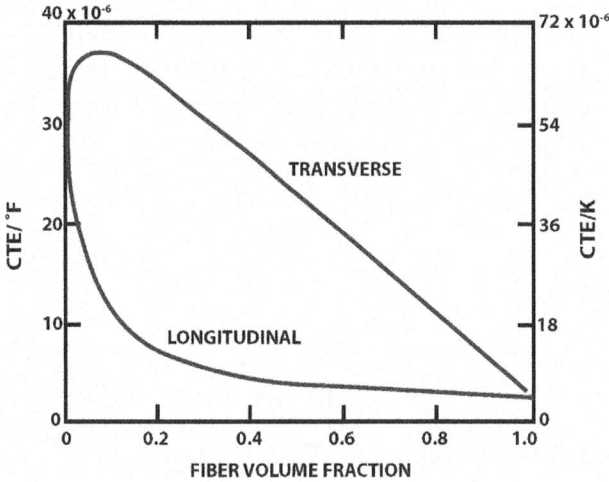

FIGURE 4.3 Variation of CTE of a unidirectional glass fibre laminate with fibre volume.

4.3 ROLE OF THE MATRIX

As with the fibres, the matrix plays an important role too, as it is in charge of bonding and holding the filaments or fibres in place, protecting filaments from the environment since it has excellent chemical resistance and dimensional stability, improving impact and fracture resistance, and providing transverse strength.

It has a very good modulus of elasticity, which allows it to act as a load transfer medium between the fibres, enabling the fibres to support the majority of the load.

The matrix helps to avoid the propagation of crack growth and provides durability to the final material.

The desired properties of the matrix also include reducing moisture absorption, low shrinkage, low coefficient of thermal expansion, good flow distribution, allowing it to penetrate the fibre bundles completely and supress voids during the compacting/curing process.

4.3.1 BONDS AND HOLDS THE FILAMENTS OR FIBRES IN PLACE

The matrix can support the fibres, keeping them in their correct position, combined to conform to any complex geometry. Fibres by themselves can be placed in any complex shape in different ways, but the matrix will be the one to keep them together in that position. Figure 4.4 is a simple representation of what was mentioned, showing the fibres and the matrix separately first, and as a result, a combined composite material.

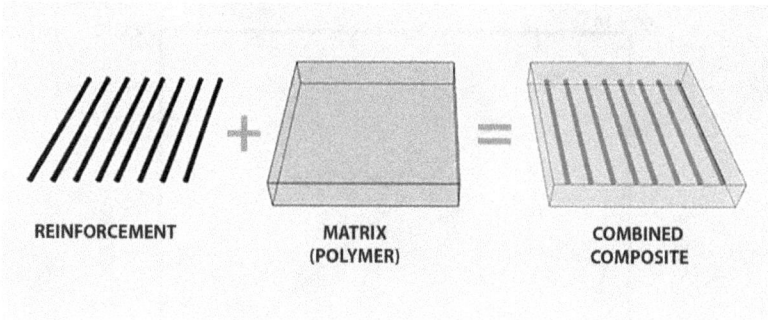

REINFORCEMENT   MATRIX   COMBINED
                          (POLYMER)   COMPOSITE

FIGURE 4.4   Representation of a combined composite material.

4.3.2   Protects Filaments from the Environment

Polymer matrix composites (PMC) are frequently considered to be immune to environmental effects. Of course, there are some limitations, but it is the matrix that is in charge of keeping the reinforcing filaments safe.

The molecules in a polymeric material occupy most of the volume. However, the space between molecules can be a significant fraction of the total volume, allowing the large molecules to be absorbed and diffuse throughout the material.

4.3.3   Improves Impact and Fracture Resistance

The resin toughness plays a dominant role in the interlaminar fracture of composite materials. In Figure 4.5, an interlaminar crack growth is occurring through the laminate due to a potential delamination, which could be induced by cyclic ply stress in the composite laminate. The delamination fracture toughness will be enhanced by increasing the ductility and decreasing the yield strength of the matrix resin.

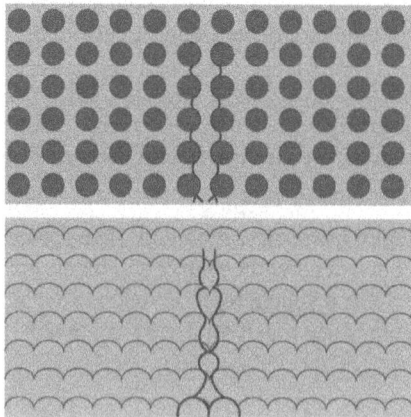

FIGURE 4.5   Interlaminar crack branching.

4.3.4 PROVIDES TRANSVERSE AND THROUGH-THE-THICKNESS STRENGTH

The matrix acts as a load transfer medium, distributing the loads evenly between fibres so that all fibres are subjected to the same amount of strain. As shown in Figure 4.6, when composites are exposed to a transverse loading condition, there will be some matrix regions in parallel with fibres and others in series with fibres. The total transverse strain will be the result of the strain in the fibres and the matrix.

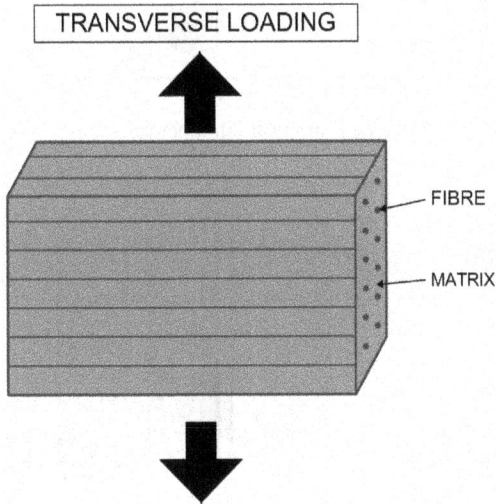

FIGURE 4.6   Representation of a composite material exposed to a transverse loading condition, where some matrix regions are aligned with fibres and others are in parallel with fibres.

The stress acting on the reinforcement is equal to the stress acting on the matrix: equal stress.

4.3.5 AVOIDS PROPAGATION OF CRACK GROWTH

During a low or high velocity impact, several damage modes could occur, including matrix crack, delamination, fibre crack, and fibre pull-out. All of these damage modes are dependent on the impact parameters of both the matrix and the fibres.

Matrix cracking is a transverse cracking when 90° plies are applied. Figure 4.7 shows a comparison between a matrix crack and a composite (fibre and matrix) crack, where the growth propagates through the matrix bridging the fibres.

Debonding at the fibre matrix interface, fibre and matrix fracture, and fibre pull-out are involved in the propagation of cracks in composite laminates. Figure 4.8 presents a schematic of these processes, using a single filament of fibre. To explain this sequence, as the load increases, the polymer matrix fails first (a) due to its significantly lower strength compared to the fibre. When a crack reaches the fibre, it remains intact because of its high strength. However, as the load continues to increase, failure

CRACK ON MATRIX ONLY	FIBRE + MATRIX CRACK

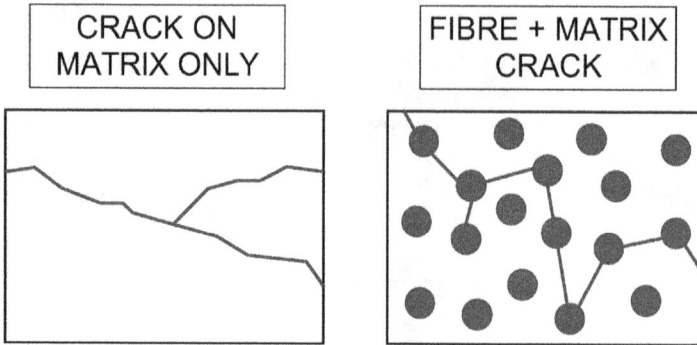

FIGURE 4.7 Comparison of crack propagation when it is only in a matrix and when it occurs in a composite.

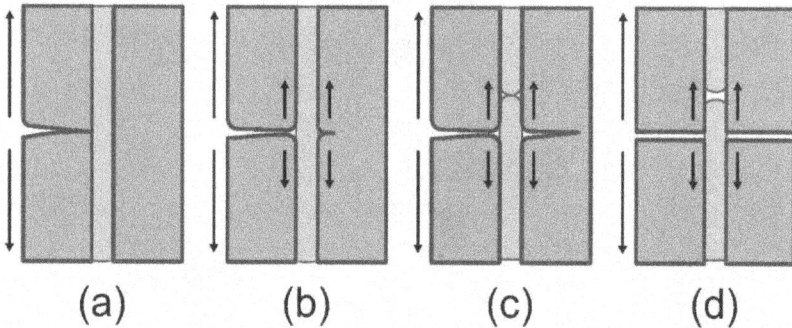

FIGURE 4.8 Fibre-toughening mechanism.

occurs at the interface between the fibre and the matrix (b), which is typically weaker than the fibre itself, allowing the crack to spread beyond the fibre. Eventually, the fibre will break at some point, leading to a visible fracture in the composite (c). Since the fibre's breakage may not align with where the matrix crack is located, fibre pull-out (d) would occur.

It is very important to correctly manage the interface structure between the matrix and the reinforcing fibres, as well as the overall structure of the matrix and the choice of fibre fillers.

4.3.6 Provides Durability

There is always an interface between the constituent materials in composites. For better results and long-term durability, the phases must bond where the interface joint occurs. Figure 4.9 shows a representation of interface adhesion between the primary matrix phase and the secondary phase reinforcement in a composite material, where the reinforcement phase is combined within a matrix by the 'Interface' (bonding region between the two phases).

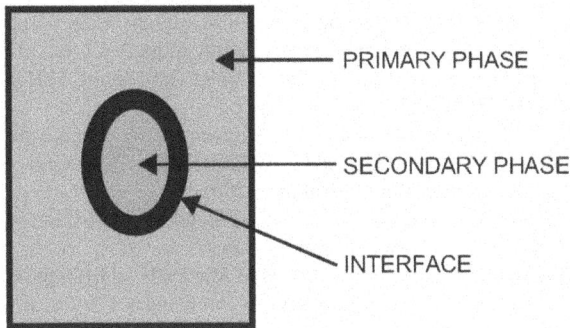

FIGURE 4.9 Interface adhesion between matrix and reinforcement.

4.4 KEY POINTS IN THE MATRIX SELECTION

The selection of the matrix is driven by the composite's application, which depends on many factors, including resistance to environmental agents, the operating conditions, mechanical properties, and cost. However, there are also other key points to consider when selecting the matrix.

The choice of a matrix could be dictated, for example, by continuously or discontinuously reinforced fibres:

- Continuous Fibres: transfers the load to the reinforcing fibres. As a result, the composite strength will be led by the fibre strength. The role of the matrix is to provide efficient transfer of load to the fibres and blunt cracks in case of fibre failure.
- Discontinuous Reinforced Composites: in this case, the matrix may lead the composite strength. The choice of the matrix will be purely affected by the required composite strength, and a higher strength matrix may be required.

The choice of the matrix also includes potential reinforcement/matrix reactions, thermal stresses (thermal mismatch between reinforcements and matrix), and matrix fatigue behaviour under cyclic conditions. A thermal mismatch occurs when a large melting temperature difference may result in matrix creep while the reinforcements remain elastic, even at temperatures approaching the matrix melting point.

The volume of resin to be used is also a key point to consider. When too much resin is used, the part is classified as "Resin Rich". On the other hand, if there is too little resin, the part is called "Resin Starved".

A resin-rich part is more prone to cracking due to the lack of fibre support. On the other hand, a resin-starved part is weaker because of void areas and the fact that fibres are not held together, which means they are not well supported.

BIBLIOGRAPHY

'Analysis and Performance of Fiber Composites' (Third edition). Bhagwan D. Agarwal, Lawrence J. Broutman and K. Chandrashekhara. Wiley India Pvt. Ltd., 2015.

'Composite Materials: Design and Applications'. Daniel Gay, Suong V. Hoa and Stephen W. Tsai. CRC Press LLC, 2003.

'Composite Materials Handbook. Volume I – Polymer Matrix Composites Guidelines for Characterization of Structural Materials – MIL-HDBK-17'. McGraw Hill/Departments and Agencies of the Department of Defence, 2002.

'Engineering Mechanics of Composite Materials' (Second edition). Isaac M. Daniel and Ori Ishai. New York: Oxford University Press, 2006.

'Fiber Composite Analysis and Design: Composite Materials and Laminates. Volume I'. Z. Hashin, B. W. Rosen, E. A. Humphreys, C. Newton and S. Chaterjee. Washington, DC: U.S. Department of Transportation: Federal Aviation Administration Office of Aviation Research, 1997.

'Handbook of Composites' (Second edition). S. T. Peters. Mountain View, CA, USA: Process Research, 1997.

'Marine Composites' (Second edition). Eric Green Associates. MD: Eric Green Associates Inc., 1999.

5 Materials Categories

When it comes to materials, there are two major categories that should be considered:

- Alloys
- Composites.

While both are made up of combinations of different substances, they have distinct differences that make them unique, and it is important to understand these differences.

5.1 WHAT ARE ALLOYS?

An alloy is a homogeneous material that can be made by melting two or more elements, one of which is a metal.

The participating metal is called the parent metal, which serves as the base metal for the alloying element and is considered an alloying agent for most substances. The alloying agent can be a non-metal or metal in which the contribution or proportion is very small.

Alloys have completely new characteristics, different from their components, and each alloy has characteristics based on the combination of elements used to create it, such as:

- More favourable mechanical properties: low density and high strength
- Increased corrosion resistance: resistance to abrasion and corrosion
- Colour change: shiny due to the presence of a metallic component in their composition.
- Present more desirable properties than those of the constituents.

An alloy can be in the type of a substitutional alloy or an interstitial alloy. Interstitial alloys contain smaller atoms that occupy the spaces between larger atoms, while substitutional alloys consist of atoms with relatively similar sizes, in which some of the atoms in the metallic crystals are replaced by atoms of another component.

Examples of substitutional alloys include bronze and brass (Figure 5.1 a), while on the interstitial alloys, we could find steel, which is made by adding carbon to iron (Figure 5.1 b).

Examples of Alloys:

- Steel –an alloy of iron and carbon
- Brass –an alloy of zinc and copper
- Bronze –an alloy of copper and tin
- Nichrome –an alloy of nickel, chromium, and iron
- Amalgam –an alloy of mercury and sodium

DOI: 10.1201/9781003565222-5

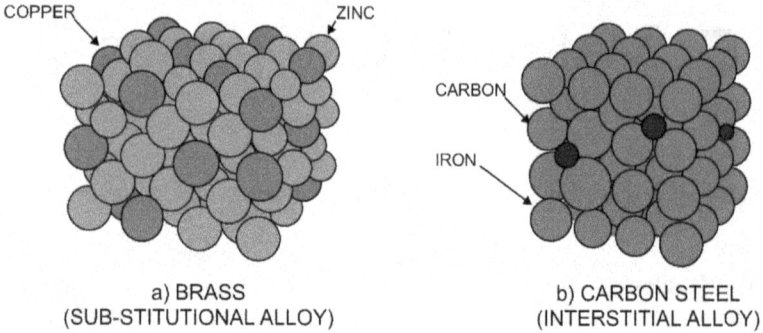

FIGURE 5.1 Atomic configuration in a substitutional alloy and an interstitial alloy.

5.1.1 Uses of Alloys

Some of the most common uses of alloys are as follows:

- Brass: Used mostly to make kitchenware, screws, locks, door handles, electrical appliances, zippers, musical instruments, decorations, gifts, etc.
- Bronze: Used to make statues, coins, medals, kitchen utensils, among other things.
- Sterling Silver: Used in surgical equipment, musical instruments, cutlery, and jewellery.
- Stainless Steel: Used to build railways, bridges, exterior features, etc. Stainless steel is very good at resisting corrosion brought about by oxidation, the process that creates rust on normal steel. Most boat deck fittings are made of stainless steel, including railings, metal handrails, pushpits, pulpits, ladders, snapper racks, and anchor rods.
- Aluminium Alloys: Used in the construction of fuselages and their components as they are lightweight.

5.2 WHAT ARE COMPOSITES?

Composites are usually a combination of two or more elements, but their structure does not include metals. Composites are heterogeneous, unlike alloys.

Composites are made up of components with different physical properties, among which the most important ones can be listed as follows:

- High strength
- High modulus
- Low density
- Excellent resistance to fatigue, creep rupture, and corrosion
- Low coefficient of thermal expansion

Composites can be made from any combination of two or more materials as long as they can be fused together into one cohesive fabric. Common examples include fibreglass (glass fibres + resin) or even concrete (cement + gravel + sand). Other examples include carbon-fibre-reinforced plastic (plastic + carbon fibres) and wood-plastic composite (wood particles + plastic).

The properties of each composite depend on the combination used to create it – for example, concrete has great compressive strength, while carbon-fibre-reinforced plastic has excellent tensile strength.

5.2.1 Uses of Composites

Composite materials have a wide range of applications in both daily life and various industries. Some of the most common applications of composites in everyday life are mainly in the naval industry, such as boat hulls and decks. However, they are also used in other fields such as construction, buildings, bridges, tanks, and automotives.

Some of the most common examples in the industry and daily life are follows:

Industry

- Aerospace: Composites are extensively used in aircraft structures, including wings and fuselage, due to their high strength-to-weight ratio.
- Marine: Boat hulls and other marine components are made from composites to enhance performance and resistance to corrosion.
- Automotive: In addition to consumer vehicles, composites are used in high-performance and electric vehicles for weight reduction and improved efficiency.
- Construction: Composites are utilised in bridges, beams, and other structures for their strength, durability, and resistance to environmental factors.
- Medical Devices: Many medical instruments and implants use composite materials for their biocompatibility and strength.
- Wind Energy: The blades of wind turbines are often made from composites to maximise efficiency and withstand environmental stresses.

Daily life

- Sports Equipment: Many items like tennis rackets, bicycles, and golf clubs use composites for enhanced strength and reduced weight.
- Consumer Electronics: Products such as smartphones and laptops often incorporate composite materials for better durability and lightweight design.
- Automobiles: Composite materials are used in components like body panels and interiors to improve fuel efficiency and reduce weight.
- Home Construction: Composites are found in flooring, roofing materials, and insulation, offering durability and thermal efficiency.

These applications highlight the versatility and advantages of composite materials across various sectors.

5.3 ALLOYS VS COMPOSITES

The main difference between an alloy and a composite is that an alloy combines two or more elements in a single substance (metal/metal or metal/non-metal). In contrast, a composite combines two or more substances into one material. For example, steel is an alloy composed of iron and carbon. On the other hand, fibreglass is a composite that combines glass fibres with resin to form a strong material.

- Alloys can be created through various methods, including casting, while composites can be made through multiple methods, including lamination and moulding.
- Alloys are typically less expensive to produce than composites, while composites are generally more expensive than alloys.
- Alloys are typically heavier than composites, while composites are generally lighter than alloys.
- Alloys are often used in applications where strength is important, while composites are often used in applications where weight is important.
- Alloys can be recycled, while composites cannot be easily recycled.
- Alloys can corrode, while composites do not rust.

Strength is one property that increases with alloying elements. Other properties may decrease or stay the same. Table 5.1 resumes the key differences between composites and alloys in terms of properties.

TABLE 5.1
Alloys vs composites key differences.

Property	Alloys	Composite Materials
Strength	Generally high and uniform	High, but varies based on fibre orientation and matrix
Ductility	Often ductile, depending on composition	Generally less ductile; more brittle in some cases
Weight	Heavier compared to composites	Lighter due to lower density of components
Fatigue Resistance	Good, but can vary	Excellent, particularly with proper design
Manufacturing Process	Usually simpler and more standardised	Often more complex, involving layering or moulding
Cost	Generally lower than advanced composites	Can be higher due to materials and processing

5.3.1 Tensile Strength in Composite Materials vs Alloys

The tensile strength of composite materials and alloys can vary significantly, as shown in Figure 5.2. These variations are based on their composition, structure, and specific applications.

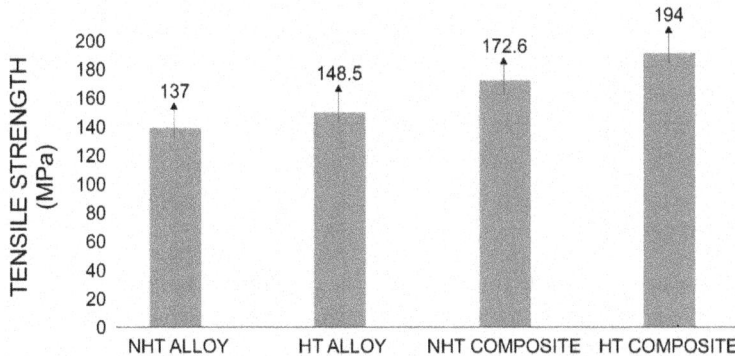

FIGURE 5.2 Tensile strength comparison between alloys and composites.

Alloys often have uniform tensile strength, which is consistent throughout the material, and composites can achieve high tensile strengths due to the combination of materials, but their strength can vary based on fibre orientation, matrix material, and manufacturing processes.

BIBLIOGRAPHY

'Analysis and Performance of Fiber Composites' (Third edition). Bhagwan D. Agarwal, Lawrence J. Broutman and K. Chandrashekhara. Wiley India Pvt. Ltd., 2015.
'Fiber Composite Analysis and Design: Composite Materials and Laminates. Volume I'. Z. Hashin, B. W. Rosen, E. A. Humphreys, C. Newton and S. Chaterjee. Washington, DC: U.S. Department of Transportation: Federal Aviation Administration Office of Aviation Research, 1997.
'Marine Composites' (Second edition). Eric Green Associates. MD: Eric Green Associates Inc., 1999.
'Mechanics of Composite Materials' (Second edition). Robert M. Jones. Taylor & Francis, 1999.
'Shigley's Mechanical Engineering Design' (Tenth edition). Richard G. Budynas and J. Keith Nisbett. McGraw Hill Education, 2015.
'Structural Composite Materials'. F. C. Campbell. ASM International, 2010.

6 Material Types

Materials can be classified based on their mechanical properties and how they respond to stress and strain in different directions. Understanding these classifications is essential in materials science and engineering, as it influences the selection and application of materials in various fields.

The different material types based on their properties are: isotropic, anisotropic, and orthotropic.

6.1 ISOTROPIC

Isotropic materials are well known for their main characteristic: they have the same material properties in all directions. We could translate this as follows: despite the direction in which the stress or load is applied, the material will react in the same way in each direction, exhibiting equal properties in every direction. Common examples include metals like steel and aluminium, which are often used in structural applications due to their consistent strength and ductility.

6.2 ANISOTROPIC

Anisotropic materials, on the other hand, have different material properties depending on the direction of the applied load. This means that their strength, stiffness, and other mechanical characteristics can differ based on the orientation of the material.

Anisotropic behaviour is commonly found in composites, wood, and certain crystals, where the arrangement of fibres or grains significantly affects performance. In this kind of material, it is hard to measure or predict their properties.

6.3 ORTHOTROPIC

Orthotropic materials are a special case of anisotropic materials having unique properties along the three mutually perpendicular axes, resulting in different mechanical characteristics in each direction. This behaviour is often seen in materials such as fibre-reinforced composites, where the fibres provide strength in one direction, while the matrix contributes to other properties in the orthogonal directions.

Even though we can still say that they are measurable materials and their properties can be predicted, presenting unique challenges in practice.

Thinking of the orthotropic material as a 2D representation of a composite lay-up or stack of plies forming a thin plate, a state of plane stress can be assumed:

- The through-thickness stress is zero
- Out-of-plane share stress is zero

DOI: 10.1201/9781003565222-6

Figure 6.1 is a representation of a composite coordinate system definition, where

- Fibre = Direction 1
- Transverse to fibre (in-plane) = Direction 2
- Normal to fibre (through-thickness) = Direction 3

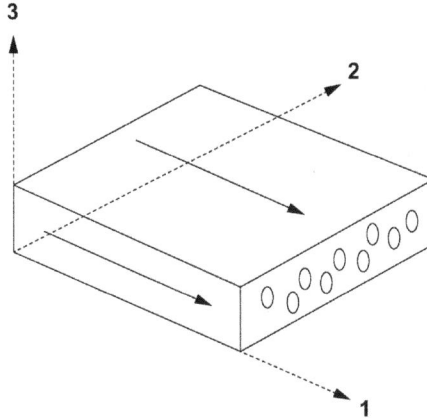

FIGURE 6.1 Representation of a composite coordinate system.

BIBLIOGRAPHY

'Analysis and Performance of Fiber Composites' (Third edition). Bhagwan D. Agarwal, Lawrence J. Broutman and K. Chandrashekhara. Wiley India Pvt. Ltd., 2015.

'Composite Materials: Design and Applications'. Daniel Gay, Suong V. Hoa and Stephen W. Tsai. CRC Press LLC, 2003.

'Composite Materials Handbook. Volume I – Polymer Matrix Composites Guidelines for Characterization of Structural Materials – MIL-HDBK-17'. McGraw Hill/Departments and Agencies of the Department of Defence, 2002.

'Engineering Mechanics of Composite Materials' (Second edition). Isaac M. Daniel and Ori Ishai. New York: Oxford University Press, 2006.

'Fiber Composite Analysis and Design: Composite Materials and Laminates. Volume I'. Z. Hashin, B. W. Rosen, E. A. Humphreys, C. Newton and S. Chaterjee. Washington, DC: U.S. Department of Transportation: Federal Aviation Administration Office of Aviation Research, 1997.

'Handbook of Composites' (Second edition). S. T. Peters. Mountain View, CA, USA: Process Research, 1997.

'Introduction to Composite Materials Design'. Ever J. Barbero. USA: Department of Mechanical & Aerospace Engineering – West Virginia University/Taylor & Francis, 1998.

'Laminated Composite Plates'. David Roylance. Cambridge, MA: Department of Materials Science and Engineering Massachusetts Institute of Technology, 2000.

'Mechanics of Composite Materials' (Second edition). Robert M. Jones. Taylor & Francis, 1999.

'Structural Composite Materials'. F. C. Campbell. ASM International, 2010.

7 Composite Material Application Field

This chapter highlights the increasing adoption of composites in sectors such as automotive, aerospace, civil engineering, medical devices, naval applications, wind energy, and sports equipment.

The list below attempts to visualize the ramifications of composite materials across different fields.

- Automotive industry
- Aerospace and military aircraft
- Commercial aircraft
- Helicopters
- Civil engineering
- Medical devices
- Naval industry
- Wind turbines
- Sport equipment

7.1 AUTOMOTIVE INDUSTRY

Composite materials are propagating rapidly in the automotive industry. Experts says that a high volume of advanced composite materials could be used more often in the future. This is related to the high demand and how competitive the industry is.

The main motivation for introducing Polymer Matrix Composites is cost savings. Like most other industries, the automotive industry continues to be interested in saving weight to reach the goal of fuel efficiency. Because of the lightweight, fuel costs will be reduced.

- Reduced weight = Fuel efficiency

Another potential technical advantage of polymer matrix composites is the corrosion resistance. Because of improved corrosion resistance, PMC automobiles could last 20 or more years compared to the current average vehicle lifetime. Advanced composites also offer substantial improvements in specific mechanical properties, with the possibility of reducing weight while increasing strength and stiffness. Figure 7.1 illustrates what a full carbon fibre car body would look like.

Finally, because they do not rust, PMCs offer significantly improved corrosion resistance over steel or galvanized steel.

While composite materials offer numerous advantages, challenges remain in terms of manufacturing costs, recycling, and integration with traditional materials.

DOI: 10.1201/9781003565222-7

FIGURE 7.1   Car body made of composite material.

The automotive industry is actively researching ways to reduce costs and improve the recyclability of composites, aiming for more sustainable practices.

In conclusion, composite materials are transforming the automotive industry by providing solutions that enhance vehicle performance, safety, and sustainability. As technology advances and manufacturing processes improve, the role of composites is expected to expand, further shaping the future of automotive design and engineering.

7.2   AEROSPACE AND MILITARY AIRCRAFT

Composite materials have revolutionised the aerospace industry, particularly in the design and manufacture of military aircraft. Their unique properties contribute to improved performance, durability, and efficiency, making them essential for modern aerospace applications. Figure 7.2 illustrates what a full carbon fibre military aircraft would look like.

Advanced Composites are used extensively today in small military aircraft, commercial aircraft, and prototype business aircraft. But the market is fully focused on applying advanced composites in large military and commercial transport aircraft.

Epoxy resins are the main matrix material used in aerospace applications, and the most common reinforcements are carbon, aramid, and high-stiffness glass fibres. Compared with metals, the main advantages of advanced composites in aerospace applications are the result of a high specific strength and stiffness, resulting in weight savings that range between 10 and 60 percent compared with metal designs. This weight reduction can be used to increase range, payload, manoeuvrability, and speed, or even more, to reduce fuel consumption.

FIGURE 7.2   Military aircraft made of composite material.

7.3   COMMERCIAL AIRCRAFT

Composite materials are transforming the commercial aviation industry by enabling the development of lighter, more fuel-efficient, and environmentally friendly aircraft. As technology continues to advance, the role of composites is expected to expand, further enhancing the performance and sustainability of commercial aviation. Their integration into aircraft design not only improves operational efficiency but also supports the industry's commitment to reducing its environmental impact.

If aramid and glass fibre-reinforced composites are included, the volume of composites used in commercial and business aircraft is about twice that used in military aircraft. Polymer Matrix Composites could make up 65 percent of the structural weight of commercial transport aircraft.

7.4   HELICOPTERS

With the exception of the all-composite business aircraft prototypes, advanced composites have been pushed more into the helicopter industry. Military applications have led the way, and the advantages of advanced composites are very similar to those in the aircraft industry:

- Weight reduction
- Parts consolidation
- Resistance to fatigue and corrosion.

Advanced composites have become the baseline materials for rotors, blades, and tail assemblies. As shown in Figure 7.3, materials such as carbon fibre/epoxy are

FIGURE 7.3 Helicopter fuselage made of composite materials.

likely to be used in the airframe, bulkheads, tail booms, and vertical fins, while glass/epoxy PMCs could be used in the rotor systems.

7.5 CIVIL ENGINEERING

There is a high volume of PMC applications in the market of construction or civil engineering, such as

- Buildings
- Bridges
- Housing

Additional applications include construction equipment:

- Cranes
- Booms
- Outdoor drive systems

The main advantages of using PMCs in construction would be

- Reduced overall systems costs
- Reduced transportation and construction costs due to lighter weight structures
- Reduced maintenance and lifetime costs due to improved corrosion resistance

Bridges are likely to be the first large-scale construction application for PMCs in the United States. Decking materials are likely to be vinylester or epoxy resins reinforced with continuous glass fibres. Cables would probably be reinforced with graphite or aramid fibres, because of the high stiffness and low creep requirements.

7.6 MEDICAL DEVICES

Polymer matrix composite materials are currently being developed for medical prostheses and implants. A perfect example of this continuous progress are the hand prosthetics, as shown in Figure 7.4. The impact of PMCs on orthopaedic devices is growing every year. Unfortunately, medical devices are not part of an industry that is expected to provide a large volume of production for PMCs.

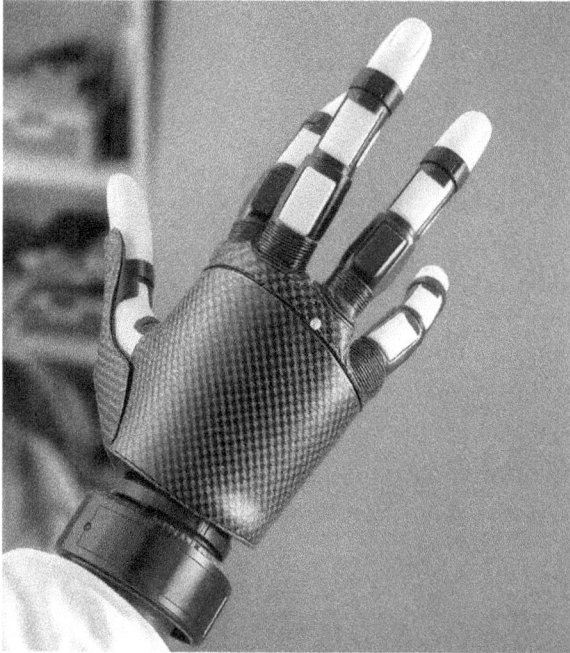

FIGURE 7.4   Composite material hand prosthesis.

- Hips
- Knees
- Bone plates
- Intramedullary nails

Metallic implant devices, such as the total hip unit that has been used since the early 1960s, suffer a variety of disadvantages:

- Difficult to fix in place
- Allergic reactions
- Poor matching of elastic stiffness
- Mechanical failure (fatigue)

7.7 NAVAL INDUSTRY

The lightweight and corrosion resistance of PMCs make them attractive for a number of naval applications. Most of their use is referred to racing boats, where specific properties and complex forms can be found.

Similar to the automotive and aircraft applications, apart from the notorious improvement in specific mechanical properties, the weight reduction combined with strength and stiffness makes them the ideal material for this application. Also, the ability not to rust and the corrosion resistance play an important role in the naval industry.

Figure 7.5 shows a generic racing sail boat as an example of the multiple composite parts and applications that can be used in this industry.

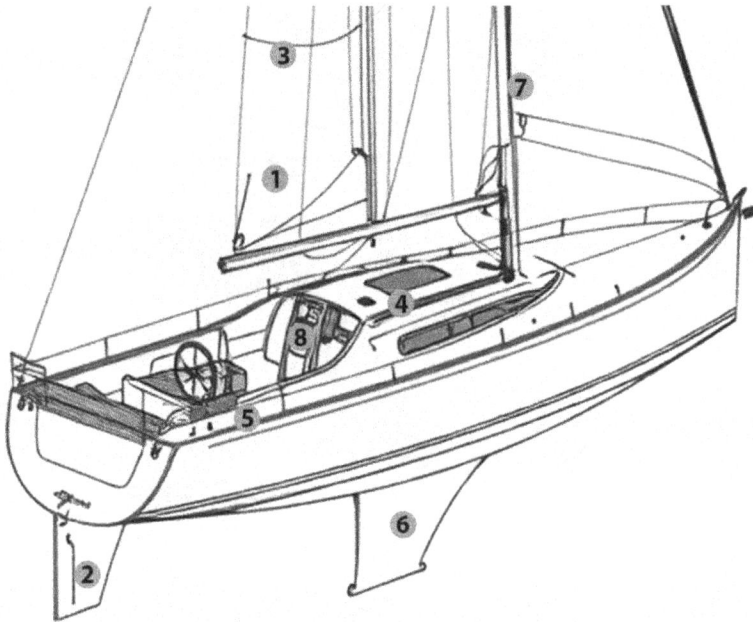

FIGURE 7.5 Racing composite sail boat with multiple composite part applications.

1- Sails: carbon fibre tow for stiffening.
2- Rudders: carbon/glass, woven/UD.
3- Sail Battens: glass/carbon prepreg.
4- Hardware: carbon fibre composites.
5- Hull and Deck: carbon/glass prepreg, woven/UD.
6- Keel: carbon/glass prepreg.
7- Mast, boom, spreaders: UD carbon tape, woven carbon, prepreg.
8- Interior fittings and bulkheads

7.8   OTHER INDUSTRIES

There are also other industries and an infinity of applications for polymer matrix composites.

- Wind turbines (Figure 7.6)
- Typical length between 25 m to 50 m
- Complex forms
- Specific properties

FIGURE 7.6   Wind turbine.

- Durability
- Sport equipment
- Bikes
- Golf
- Fishing rods
- Tennis racquets
- Hockey sticks
- Skis
- Snowboards

BIBLIOGRAPHY

'Composite Materials: Design and Applications'. Daniel Gay, Suong V. Hoa and Stephen W. Tsai. CRC Press LLC, 2003.

'Engineering Mechanics of Composite Materials' (Second edition). Isaac M. Daniel and Ori Ishai. New York: Oxford University Press, 2006.

'Handbook of Composites' (Second edition). S. T. Peters. Mountain View, CA, USA: Process Research, 1997.

'Marine Composites' (Second edition). Eric Green Associates. MD: Eric Green Associates Inc., 1999.

'Mechanics of Composite Materials' (Second edition). Robert M. Jones. Taylor & Francis, 1999.

'Theory of Composites Design'. Stephen W. Tsai. Department of Aeronautics and Astronautics: Stanford University, 1992.

8 Advantages and Disadvantages of PMCs

PMCs have become increasingly popular in various industries, including aerospace, automotive, marine, and construction, due to their versatility and performance characteristics.

Understanding the advantages and disadvantages of PMCs is essential for selecting the right materials for specific applications.

8.1 ADVANTAGES

- **Good Stiffness and Strength:** PMCs provide excellent stiffness and strength, which are critical for structural applications. The combination of a strong reinforcement (like carbon or glass fibres) with a polymer matrix allows these materials to withstand significant loads without deforming. This makes PMCs ideal for components that require both rigidity and load-bearing capacity, such as in aerospace and automotive applications.

- **Low Density:** One of the standout features of PMCs is their low density compared to traditional materials like metals. This lightweight characteristic leads to significant reductions in overall weight for structures and components, enhancing performance and fuel efficiency, particularly in transportation applications where every gram counts.

- **Low Cost:** Many types of polymer matrix composites can be produced at lower costs compared to high-performance metal alloys. With advancements in manufacturing processes, such as resin infusion and automated lay-up, the overall cost of producing PMCs has decreased, making them more accessible for a wider range of applications.

- **High Corrosion Resistance:** PMCs exhibit excellent resistance to corrosion, particularly when exposed to harsh environmental conditions, such as moisture and chemicals. This property extends the lifespan of components and reduces maintenance costs, making PMCs suitable for marine, automotive, and industrial applications.

- **Low Thermal Expansion:** The coefficient of thermal expansion (CTE) for many PMCs is relatively low, meaning that they do not expand or contract significantly with changes in temperature. This stability is crucial for applications where dimensional accuracy is important, such as in aerospace components and precision instruments.

- **Excellent In-Service Experience:** PMCs have demonstrated reliable in-service performance across various industries. Their durability and ability

    DOI: 10.1201/9781003565222-8

to withstand cyclic loading, environmental exposure, and other operational stresses contribute to a positive track record, instilling confidence in their use for critical applications.

- **Light in Weight:** The lightweight nature of PMCs is one of their most significant advantages. This property not only enhances fuel efficiency in vehicles and aircraft but also improves the overall performance and manoeuvrability of structures. In applications like sporting goods and consumer products, lower weight can enhance usability and comfort.

- **High Strength-to-Weight Ratio:** They offer an excellent strength-to-weight ratio, making them ideal for applications that require both strength and reduced mass.

- **Design Flexibility:** The manufacturing processes for PMCs allow for complex shapes and structures, enabling innovative designs that can optimise performance and aesthetics.

8.2   DISADVANTAGES

- **Low Maximum Working Temperature:** PMCs often have a lower maximum working temperature compared to metals and some other materials. Many polymer matrices can degrade or lose their mechanical properties at elevated temperatures, limiting their use in high-temperature applications. This can be a critical drawback in industries like aerospace, where components may be subjected to extreme heat.

- **Processing Temperatures Are Higher:** The manufacturing processes for PMCs typically require higher processing temperatures, especially for thermoset resins. This can increase energy consumption and lead to longer production times. Additionally, maintaining these temperatures can complicate the manufacturing process, necessitating more robust equipment and monitoring.

- **Require Special Processing Equipment:** The production of PMCs often requires specialised equipment and techniques, such as autoclaves for curing or vacuum bagging for resin infusion. This need for specific machinery can increase initial investment costs and complicate the manufacturing process, making it less accessible for smaller manufacturers or less specialised applications.

- **Thermoset Resins Have Poor Impact Resistance:** While PMCs offer many mechanical advantages, thermoset resins, which are commonly used in composite applications, tend to have poor impact resistance. This brittleness can lead to cracking or failure under high-impact conditions, making them less suitable for applications where toughness is critical, such as in certain structural or safety components.

- **Environmental Concerns:** The disposal and recycling of PMCs can be problematic due to the difficulty of separating the matrix from the fibres, raising environmental concerns.

- **Brittleness:** Depending on the matrix and reinforcement used, some PMCs can be brittle and susceptible to cracking under certain conditions.

BIBLIOGRAPHY

'Analysis and Performance of Fiber Composites' (Third edition). Bhagwan D. Agarwal, Lawrence J. Broutman and K. Chandrashekhara. Wiley India Pvt. Ltd., 2015.

'Composite Materials: Design and Applications'. Daniel Gay, Suong V. Hoa and Stephen W. Tsai. CRC Press LLC, 2003.

'Composite Materials Handbook. Volume I – Polymer Matrix Composites Guidelines for Characterization of Structural Materials – MIL-HDBK-17'. McGraw Hill/Departments and Agencies of the Department of Defence, 2002.

'Engineering Mechanics of Composite Materials' (Second edition). Isaac M. Daniel and Ori Ishai. New York: Oxford University Press, 2006.

'Fiber Composite Analysis and Design: Composite Materials and Laminates. Volume I'. Z. Hashin, B. W. Rosen, E. A. Humphreys, C. Newton and S. Chaterjee. Washington, DC: U.S. Department of Transportation: Federal Aviation Administration Office of Aviation Research, 1997.

'Handbook of Composites' (Second edition). S. T. Peters. Mountain View, CA, USA: Process Research, 1997.

'Introduction to Composite Materials Design'. Ever J. Barbero. USA: Department of Mechanical & Aerospace Engineering – West Virginia University/Taylor & Francis, 1998.

'Theory of Composites Design'. Stephen W. Tsai. Department of Aeronautics and Astronautics: Stanford University, 1992.

9 Materials Used in the Naval Industry

The reinforcing materials used to make the composite material are known as fibres. These fibres are a discontinuous (or dispersed) phase that is added to the matrix to provide the properties that the matrix itself does not possess. To be clear, this component is what provides mechanical resistance, rigidity, and hardness to the material.

The most commonly used fibre reinforcing materials in the naval industry are as follows:

- Fibreglass.
- Carbon fibre.
- Aramid.

9.1 FIBREGLASS

Fibreglass, also known as glass-reinforced plastic (GRP) or glass fibre, is a lightweight and durable material made from extremely fine strands of glass woven together. Its unique properties, such as high strength, corrosion resistance, and versatility, make it ideal for use across a range of industries.

The production of fibreglass involves pulling molten glass through fine nozzles, creating thin strands known as glass fibres. These fibres are then grouped into mats, woven into cloth, or arranged into other forms, depending on the intended application.

To understand how they work and how they have been created, it is easy to think of them as a "fragile" material. When these fragile materials are exposed to stresses or tension, the existing random defects in the solid may cause a failure even at lower numbers than their theoretical values. To solve this problem, these materials are produced in the form of fibres or filaments.

The randomly oriented defects will still exist, but they will appear only in some of the thousands of filaments that comprise the reinforcement, while the rest can perform with the expected resistance of the material without defects. Figure 9.1 shows a woven fibreglass cloth made by fine plain weave pattern.

The fibres are combined with a resin, usually polyester, vinylester, or epoxy, to form a strong composite material.

It is currently the most used reinforcement in the manufacture of composite materials; this is due to its good mechanical characteristics, ease of procurement, and low cost, especially compared with carbon fibre or aramid. All these factors contribute to it being the main material in the construction of most small ships. The most common matrices used with fibreglass are polyester resins.

As an extra detail, density and tensile properties are similar to those of carbon fibre and aramid, but the strength and tensile modulus are significantly lower.

DOI: 10.1201/9781003565222-9

FIGURE 9.1   Woven fibreglass cloth.

However, fibres can perform a high stress-resisting ability only in the direction of the fibres. The same happens, for example, with filaments in a rope.

9.1.1 FIBREGLASS TYPES

Some of the most common types of fibreglass are as follows:

- **Glass A:** It has a high silica content, is sensitive to humidity, but has good resistance to chemical and acid solution attacks.
- **Glass B:** It has high durability and good electrical properties.
- **Glass C (Chemical):** Offers high resistance against chemical agents. Due to their high chemical resistance, they are usually used in the superficial layers of the laminate. It presents mechanical properties between glass A and E.
- **Glass D (Dielectric):** Due to its dielectric properties (very low electrical losses), it is used for electronic and telecommunications components.
- **Glass E (Electrical):** It is the type of fibreglass most used in composite materials. Regarding its properties, it presents high density, great resistance to humidity, good electrical properties, and good rigidity and resistance characteristics. It is widely used in naval, aeronautical, automotive applications, etc.
- **Glass R or S (Resistance):** This type of fibre is used in structures with high mechanical characteristics; this is because it has a tensile strength and a modulus of elasticity much higher than others.

9.1.2 KEY PROPERTIES OF FIBREGLASS

Lightweight. Fibreglass is much lighter than metals, making it useful in applications where weight reduction is critical.

TABLE 9.1
Properties or characteristics for the different types of fibreglass.

Letter Designation	Property or Characteristics
E, Electrical	Low electrical conductivity
S, Strength	High strength
C, Chemical	High chemical durability
M, Modulus	High stiffness
A, Alkali	High alkali or soda lime glass
D, Dielectric	Low dielectric constant

TABLE 9.2
Average property values for the common types of fibreglass.

Type	Glass-A	Glass-E	Glass-S	Glass-R
Filament diameter (μm)	5–13	10–20	10	10
Density (Kg/m^3)	2500	2580	2480	2590
Modulus of elasticity (GPa)	69	72.2	86	85
Resistance to traction (GPa)	3.1	3.4	4.59	3.4–4.4
Specific modulus	28	28	34	33
Coefficient of thermal expansion (10^{-6}/°C)	8.6	5	5.1	5

- **High fibre-matrix bonding**
- **High Strength-to-Weight Ratio:** It offers exceptional strength without compromising weight, giving it an edge over traditional materials.
- **Excellent mechanical resistance (tensile strength/density)**
- **Corrosion Resistance:** It does not rust or degrade from exposure to moisture or chemicals.
- **Low coefficient of expansion**
- **Resistant to chemical attacks**
- **Electrical Insulation:** Fibreglass is non-conductive, making it suitable for electrical and electronic applications.
- **Low thermal conductivity**
- **Low cost**

9.1.3 COMMON APPLICATIONS OF FIBREGLASS

Fibreglass is used across many industries because of its unique combination of strength, lightweight, and corrosion resistance. For example,

9.1.3.1   Automotive and Marine Industry

- Car Parts: Body panels, hoods, and bumpers due to their lightweight and impact resistance.
- Boat Hulls and Decks: Fibreglass resists water and corrosion, making it ideal for yachts, fishing boats, and kayaks.
- Truck and Trailer Bodies: Reduces weight to improve fuel efficiency.

9.1.3.2   Construction and Architecture

- Roofing Panels and Sheets: Provides durability and weather resistance for homes and industrial buildings.
- Insulation: Fibreglass wool is widely used in thermal and acoustic insulation.
- Pipes and Tanks: Storage tanks for chemicals, water, or waste rely on fibreglass to prevent corrosion.
- Cladding and Facades: Used for decorative elements, domes, and architectural features.

9.1.3.3   Aerospace and Aviation

- Aircraft Interiors: Components like cabin panels, seats, and overhead compartments are made from fibreglass to reduce weight.
- Drones and Lightweight Structures: Used in structural parts to enhance fuel efficiency and improve aerodynamics.

9.1.3.4   Sports and Recreation

- Surfboards and Snowboards: Fibreglass offers strength and flexibility, ideal for high-performance sports equipment.
- Fishing Rods: Its lightweight and resilience help create responsive rods.
- Bicycles and Helmets: Components are reinforced with fibreglass for durability and safety.

9.1.3.5   Healthcare and Medical Applications

- Medical Equipment Casings: Durable and lightweight housings for diagnostic machines.
- Orthopaedic Casts: Fibreglass is often used in modern casts due to its lightweight and durability compared to traditional plaster.

9.1.4   Advantages and Limitations of Fibreglass

9.1.4.1   Advantages

- Cost-effective compared to many other composite materials.
- Customizable for various shapes and applications.
- Low maintenance, with excellent resistance to environmental wear.

9.1.4.2 Limitations

- Brittle under extreme force, prone to cracking.
- Decomposes slowly, raising concerns about environmental sustainability.
- Fibreglass dust can cause irritation during manufacturing or handling.

Fibreglass is a versatile material that has become a staple across industries due to its adaptability, durability, and affordability. While there are some challenges in its disposal, ongoing developments aim to make fibreglass even more sustainable and widely used in the future.

9.2 CARBON FIBRE

Carbon fibre is a lightweight, strong material made from thin strands of carbon atoms tightly bonded together in a crystal alignment. These fibres are extremely thin, with a diameter of about 5 to 10 micrometres. The aligned structure provides exceptional strength-to-weight ratios, making carbon fibre a popular choice in high-performance applications.

The fibres are typically woven into fabrics or combined with resin to create carbon fibre-reinforced polymer (CFRP) composites. This composite material exhibits enhanced mechanical properties and is commonly used in industries like aerospace, automotive, and sports equipment. Figure 9.2 shows a woven carbon fibre cloth made by fine twill weave pattern.

FIGURE 9.2   Woven carbon fibre cloth.

The atomic structure of carbon fibre is similar to graphite, composed of sheets of carbon atoms organised into a regular hexagonal pattern. Figure 9.3 is a representation of the carbon fibre atomic structure with its hexagonal intersections. Graphite is the same material used in black pencil leads; for example, the difference is purely in the way the hexagonal structures intersect. Carbon fibre is a form of graphite in which these sheets are long and thin. Bundles of these ribbons are packed together to create fibres; from there is the name "carbon fibre".

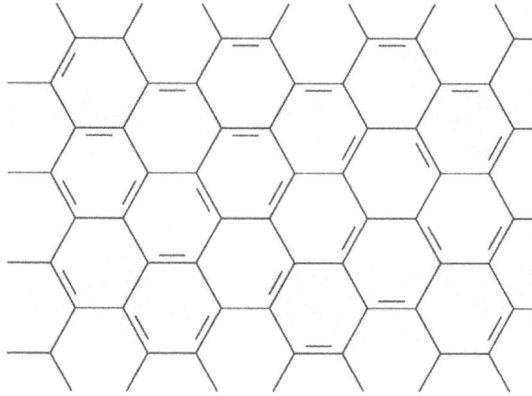

FIGURE 9.3 Carbon fibre atomic structure.

Carbon fibre is an amorphous material where the ribbons of carbon atoms are randomly packed or squeezed together. This is the reason why they "lock" within each other under tensile stress, preventing them from slipping between layers and increasing their resistance.

The carbon fibre used to reinforce composite materials is made from a polymer called "PAN" through a sophisticated process.

Carbon fibre filaments present a diameter that can vary from 5 mm to 8 mm and are combined into rovings that contain between 5,000 and 12,000 filaments.

9.2.1 CARBON FIBRE TYPES

Depending on the treatment temperature, we can find different types of carbon fibres:

9.2.1.1 Based on Modulus (Stiffness Level)

- High-modulus fibres (HM): They are the most rigid and require the highest temperature treatment; in other words, they are very stiff but expensive.

These fibres have a high modulus of elasticity, high cost, and low stretch. They are used in space or specialised high-performance components.

- High-resistance fibres (HR) or intermediate modulus: These are the strongest and carbonize at the temperature that provides the highest tensile strength, with values greater than 300 GPa. They offer a balance between strength and flexibility and are found in aerospace applications.
- Carbon fibres (HT) or standard modulus: These are the cheapest on the market. Their rigidity is less than the ones mentioned previously but they present good resistance. The treatment temperature is lower than the previous ones. They are typically used in sports goods and automotive applications.

TABLE 9.3

Average property values for the different types of carbon fibre based on modulus.

Type	High Resistance	High Modulus	Standard
Filament Diameter (μm)	8	7	7–8
Density (Kg/m³)	1740/1760	1810/1870	1820
Modulus of Elasticity (GPa)	230	390	290
Resistance to Traction (GPa)	2.6–5	2.1–2.7	3.1
Resistance to Stretch (%)	2	0.7	1.1
Specific Modulus	130	210	160
Coefficient of Thermal Expansion (10⁻⁶/°C)	2.56	2.59	2.59

9.2.1.2 Based on Precursor Material

- PAN-based (Polyacrylonitrile): The most common type, offering high tensile strength and good flexibility.
- Pitch-based: Known for excellent thermal conductivity and very high modulus but more brittle.
- Rayon-based: Primarily used in specific aerospace and defence applications for carbon-carbon composites.

9.2.1.3 Woven vs Unidirectional Fibres

- Woven Fabric: Provides strength in multiple directions; ideal for structural reinforcement.
- Unidirectional Fibre: Aligns all fibres in one direction for maximum strength along a single axis.

9.2.2 Key Properties of Carbon Fibre

The best mechanical properties of carbon fibres can be obtained by combining carbon fibres with epoxy matrices.

- **High Strength-to-Weight Ratio**: Carbon fibre is five times stronger and two times stiffer than steel while being much lighter.
- **High resistance and specific rigidity.**
- **Very low coefficient of expansion.**
- **Corrosion Resistance:** Chemically inert, making it resistant to rust and most chemicals.
- **Great dimensional stability.**
- **Good thermal conductivity.**
- **Good Resistance to Fatigue:** Can withstand repetitive stresses without significant degradation.
- **High Tensile Strength:** Excellent performance under tension, providing durability.

- **Thermal Stability:** Maintains performance under extreme temperatures but can degrade in oxidizing environments.
- **Electrical Conductivity:** Carbon fibre conducts electricity, which can be beneficial or limiting, depending on the application.

9.2.3 COMMON APPLICATIONS OF CARBON FIBRE

9.2.3.1 Aerospace and Defence
- Aircraft structures (wings, fuselages).
- Satellite components.
- UAVs (Unmanned Aerial Vehicles).

9.2.3.2 Automotive Industry
- Lightweight chassis and body panels in high-performance cars.
- Carbon fibre wheels.

9.2.3.3 Sport Equipment
- Bicycles, golf clubs, tennis rackets, and fishing rods.
- Helmets and protective gear.

9.2.3.4 Renewable Energy
- Wind turbine blades.

9.2.3.5 Medical Applications
- Prosthetics and orthotics.
- Surgical instruments and diagnostic equipment.

9.2.3.6 Construction
- Strengthening of bridges, columns, and structures through carbon fibre wraps and reinforcements.

9.2.4 ADVANTAGES AND LIMITATIONS OF CARBON FIBRE

9.2.4.1 Advantages
- High Strength with Low Weight: Ideal for applications where weight reduction without compromising strength is essential.
- Corrosion Resistance: Long-lasting performance even in harsh environments.
- Design Flexibility: Can be moulded into complex shapes and structures.
- Fatigue Resistance: Suitable for dynamic, high-stress environments.
- Aesthetic Appeal: Carbon fibre's distinct weave is visually attractive and associated with high-performance products.

9.2.4.2 Limitations
- High Cost: Production and raw materials are expensive, limiting its use to specialised applications.

- Brittleness: While strong under tension, it can fracture or crack under impact or excessive compression.
- Low impact resistance.
- Limited Recycling: Carbon fibre recycling is challenging and energy-intensive.
- Conductivity Issues: In some situations (e.g., electrical insulation), its conductivity can be a drawback.
- Environmental Impact: Production involves energy-intensive processes and some harmful chemicals.

9.3   ARAMID FIBRES (ORGANICS)

Aramid is an organic filament that comes from certain petroleum derivatives. They are used in composite structures, such as Kevlar fibres. Figure 9.4 shows a woven kevlar fibre cloth made by a fine twill weave pattern.

Kevlar is a brand name for a type of aramid fibre (aromatic polyamide) developed by DuPont in the 1960s. It is known for its exceptional strength, durability, and lightweight nature, making it widely used in various high-performance applications. Kevlar fibres are characterised by their high tensile strength, flexibility, and resistance to impact, offering outstanding protective properties.

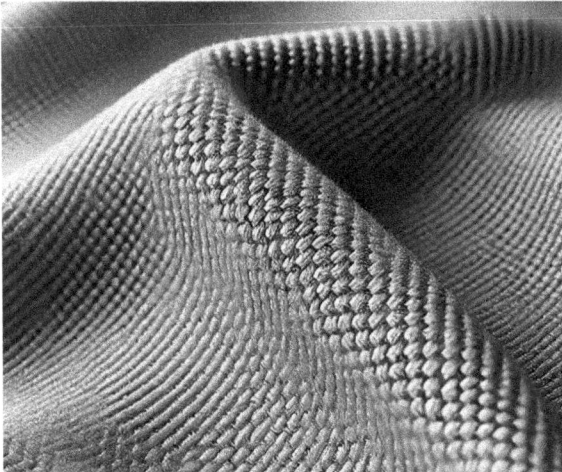

FIGURE 9.4   Woven Kevlar fibre cloth.

This type of fibre is applied in fields where high tensile strength with low weight and high impact resistance is required. Its specific traction coefficient is high and close to the carbon fibre composites, but its compressive strength is quite weak. Kevlar's molecular structure consists of strong intermolecular hydrogen bonds, giving it superior resistance to stretching and tearing compared to many other materials.

This material is used in the manufacture of boat hulls; however, it is less required compared to carbon fibre and fibreglass.

Aramid fibres have high corrosion resistance and are extremely resistant to chemical attack, except for strong acids and bases at high concentrations.

9.3.1 Aramid Fibre Types

The main aramid fibres are as follows:

- **Kevlar 29:** Known for high tensile strength, high resistance, low density, and good elastic modulus. The main purposes are ballistic protection and the manufacture of ropes and cables.

- **Kevlar 49:** Offers enhanced stiffness, high strength, high modulus, and low density. It is the most suitable type of fibre to combine with different matrices in the manufacture of composite materials. Its properties make this commercial variant suitable for composites used in aerospace, automotive, and marine applications such as boat building.

- **Kevlar 100:** Designed for high wear resistance and abrasion protection, used in gloves, cables, and protective clothing.

- **Kevlar AP (Advanced Performance):** Provides improved strength while being lighter, used in ropes, cables, and newer protective equipment.

- **Kevlar KM2:** Specially developed for military and tactical applications, including helmets and vehicle armour.

TABLE 9.4

Average property values for the different types of aramid fibres.

Type	Kevlar 29	Kevlar 49	Kevlar 100	Kevlar AP	Kevlar KM2
Filament Diameter (μm)	12	12	12	8–12	8–12
Density (Kg/m³)	1440	1450	1440	1440	1440
Modulus of Elasticity (GPa)	60	128	70–80	70–80	70–80
Resistance to Traction (GPa)	1.92	2.49–3.6	3.5	3.6	3.5
Resistance to Stretch (%)	41	88	–	–	–

9.3.2 Key Properties of Aramid Fibre

The best mechanical properties of carbon fibres can be obtained by combining carbon fibres with epoxy matrices.

- **High Specific Tensile Strength:** Kevlar is five times stronger than steel on an equal weight basis.

- **Lightweight:** It offers a high strength-to-weight ratio, making it ideal for applications requiring both strength and lightness.

- **High Impact Resistance:** Known for excellent resistance to impacts, cuts, and penetration, making it crucial in ballistic applications.

- **High modulus of elasticity**

- **Thermal Stability:** Kevlar can withstand temperatures up to 400°C (750°F) without degradation.
- **Chemical Resistance:** Resistant to many chemicals, though exposure to strong acids or UV radiation can degrade it over time.
- **Flexibility:** Despite being extremely strong, Kevlar remains flexible, making it ideal for textiles and cables.
- **High tenacity**
- **High capacity for energy absorption**

9.3.3 COMMON APPLICATIONS OF ARAMID FIBRE

9.3.3.1 Personal Protection
- Bulletproof vests, helmets, and gloves.
- Tactical and riot gear.

9.3.3.2 Aerospace and Defence
- Aircraft panels and components.
- Ballistic armour for vehicles.

9.3.3.3 Automotive
- Reinforcement of tires and brake pads.
- Composite panels in high-performance cars.

9.3.3.4 Marine Applications
- Ropes, sails, hulls and protective covers due to its resistance to abrasion and saltwater.

9.3.3.5 Industrial uses
- Protective gloves for handling sharp objects.
- High-strength ropes and cables.
- Conveyor belts and hoses.

9.3.3.6 Sport Equipment
- Reinforced materials in kayaks, racing canoes, and helmets.
- Protective clothing and gear in motorsports.

9.3.4 ADVANTAGES AND LIMITATIONS OF ARAMID FIBRE

9.3.4.1 Advantages
- Exceptional Strength: Provides reliable protection against bullets, knives, and impacts.
- Lightweight: High strength-to-weight ratio reduces load in personal gear, vehicles, and equipment.
- Thermal Resistance: Performs well under extreme temperatures, suitable for aerospace and firefighting applications.

- Flexibility and Comfort: Despite its strength, Kevlar can be woven into comfortable fabrics.

- Abrasion and Cut Resistance: Ideal for use in gloves, protective clothing, and industrial belts.

- Chemical Resistance: Withstands many chemicals, enhancing its durability in harsh environments.

9.3.4.2 Limitations

- High Cost: Production is expensive, limiting its use to specialised applications.

- Degradation from UV Light: Prolonged exposure to sunlight can degrade Kevlar fibres, requiring protective coatings or treatments.

- Brittle Under Compression: Kevlar is strong under tension but can weaken under compressive loads.

- Limited Elasticity: Not as stretchy as other materials, which can limit its flexibility in some uses.

- Difficult Recycling: Recycling Kevlar is complex and energy-intensive.

- Chemical Sensitivity: Although generally resistant, strong acids and bases can degrade it.

BIBLIOGRAPHY

'Analysis of Composite Materials. A Survey'. Z. Hashin. *Journal of Applied Mechanics*, Vol. 50 (1983), No. 3, 481–505.

'Composite Materials: Design and Applications'. Daniel Gay, Suong V. Hoa and Stephen W. Tsai. CRC Press LLC, 2003.

'Composite Materials Handbook. Volume I – Polymer Matrix Composites Guidelines for Characterization of Structural Materials – MIL-HDBK-17'. McGraw Hill/Departments and Agencies of the Department of Defence, 2002.

'Engineering Mechanics of Composite Materials' (Second edition). Isaac M. Daniel and Ori Ishai. New York: Oxford University Press, 2006.

'Fiber Composite Analysis and Design: Composite Materials and Laminates. Volume I'. Z. Hashin, B. W. Rosen, E. A. Humphreys, C. Newton and S. Chaterjee. Washington, DC: U.S. Department of Transportation: Federal Aviation Administration Office of Aviation Research, 1997.

'Handbook of Composites' (Second edition). S. T. Peters. Mountain View, CA, USA: Process Research, 1997.

'Introduction to Composite Materials Design'. Ever J. Barbero. USA: Department of Mechanical & Aerospace Engineering – West Virginia University/Taylor & Francis, 1998.

'Marine Composites' (Second edition). Eric Green Associates. MD: Eric Green Associates Inc., 1999.

'Structural Composite Materials'. F. C. Campbell. ASM International, 2010.

'Theory of Composites Design'. Stephen W. Tsai. Department of Aeronautics and Astronautics: Stanford University, 1992.

10 Reinforcing Materials Design

There are a number of factors that need to be considered when starting to design with fibre-reinforced composites. These factors will be linked directly with the requested final properties to obtain in a laminate.

- Length
- Diameter
- Orientation
- Quantity and properties of the fibres
- Matrix properties
- Bond between fibres and matrix

10.1 FIBRE LENGTH AND DIAMETER

In fibre-reinforced materials, the length and diameter of the fibres play crucial roles in determining the mechanical properties of the final composite, such as strength, stiffness, toughness, and durability. These two parameters affect how efficiently the fibres can reinforce the matrix material and influence factors like load transfer, crack resistance, and flexibility. Proper selection of fibre dimensions is essential to optimise performance for specific applications.

The fibres can be short, long, or even continuous. Their dimensions are often described by the following relation: L/d, where "L" is the length of the fibres and "d" refers to the diameter. The common fibres have diameters from 10 microns to 150 microns.

The strength of the composite improves when the length-to-diameter (L/D) ratio is large.

It is noteworthy that fibres often fracture due to surface defects. To explain this simply, as the diameter of the fibres decreases, the surface area will be less; which means there are fewer defects that could propagate during the process or even under load conditions. Long fibres are also preferred; the end of each fibre supports less load than the rest of them. Therefore, with fewer ends, the greater the ability of the fibres to carry loads.

There is a critical fibre length value (LC) where the strength and stiffness of the composite material increase. Figure 10.1 shows, for example, a "fibre length vs. tensile stiffness" graph, where the tensile resistance increases with fibre length. This critical fibre length will depend on the type of material.

DOI: 10.1201/9781003565222-10

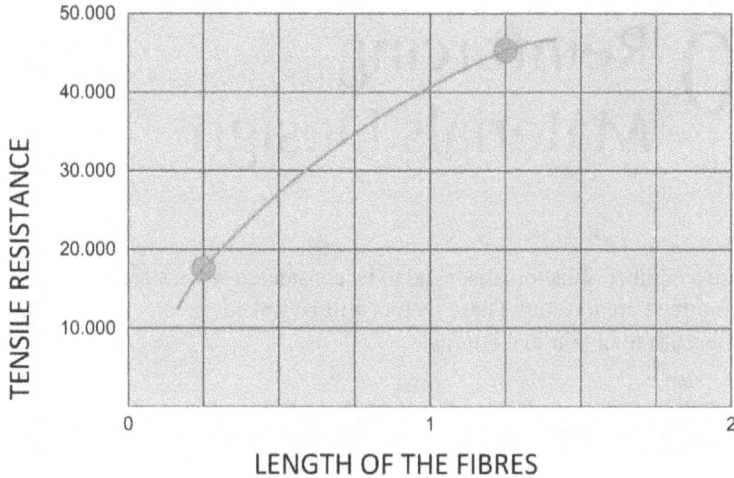

FIGURE 10.1   Fibre length vs tensile stiffness.

10.1.1 FIBRE LENGTH

- Continuous fibres: These are long, uninterrupted fibres that run through the length of the composite, offering maximum load transfer and high mechanical performance. Commonly used in aerospace, automotive, and structural applications, continuous fibres provide high strength and stiffness along the direction of alignment.

- Short (chopped) fibres: These fibres are discontinuous and randomly oriented, typically ranging from 0.1 mm to a few centimetres in length. They offer improved flexibility and ease of processing but lower mechanical performance compared to continuous fibres. They are used in injection-moulded parts and thermoplastics, where complex shapes or cost efficiency is prioritised.

- Critical Fibre Length: This is the minimum length required for a fibre to effectively transfer stress between the matrix and the fibre. If the fibre is shorter than this critical length, it may slip within the matrix, reducing reinforcement efficiency.

10.1.2 FIBRE DIAMETER

- Smaller diameters increase the surface area available for bonding with the matrix improving load transfer and the overall strength of the composite. However, very fine fibres may be more prone to damage during manufacturing or handling.

- Larger diameters are generally more durable and resistant to breakage but offer less surface area for bonding, potentially lowering the composite's mechanical performance.

10.2 FIBRE ORIENTATION

The orientation of fibres within the matrix is a key factor influencing the mechanical properties and performance of fibre-reinforced composites. Fibre orientation determines how loads are distributed throughout the material, significantly affecting properties like strength, stiffness, and impact resistance. Proper alignment of fibres ensures optimal load transfer between the matrix and the fibres, making the composite suitable for specific structural and functional applications.

The reinforcing fibres can be oriented in different ways within the matrix. Three different forms of fibre orientation can be found:

10.2.1 CONTINUOUS FIBRE COMPOSITE

Continuous fibres run uninterrupted through the length of the composite material, as shown in Figure 10.2, and they are often aligned in one or more specific directions.

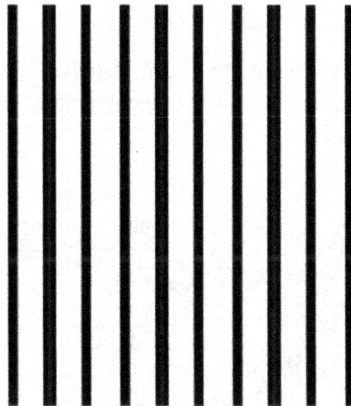

FIGURE 10.2   Representation of continuous fibre composite.

This fibre orientation offers maximum load transfer and high mechanical performance along the fibre direction.

10.2.2 DISCONTINUOUS FIBRE COMPOSITE

Discontinuous fibres are short and typically arranged in a non-continuous manner. Figure 10.3 shows a representation of a discontinuous fibre composite. They may still have some directional alignment but are limited in length.

This fibre orientation type is easier to process than continuous fibres and also has moderate mechanical strength to the previous type. They are typically used in injection moulding and compression moulding processes.

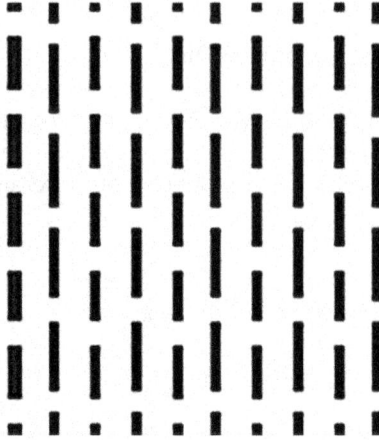

FIGURE 10.3   Representation of discontinuous fibre composite.

10.2.3   RANDOMLY ORIENTED DISCONTINUOUS FIBRES

In this configuration, discontinuous fibres are distributed randomly throughout the matrix, as shown in Figure 10.4, resulting in a composite with nearly isotropic properties (uniform strength in all directions).

FIGURE 10.4   Representation of randomly oriented discontinuous fibre composite.

This fibre orientation provides good impact resistance and crack propagation control, as well as good resistance where multi-directional loads are expected.

One of the main characteristics of fibre-reinforced composites is that their properties can be designed to withstand different loading conditions.

Long and continuous fibres can be combined into the matrix in different directions, perpendicular to each other (0°/90° layers), resulting in a good resistance in the two perpendicular directions. More complex arrangements could be displayed as well, such as 0°/±45°/90° layers, to provide reinforcement in many directions. In an ideal situation, the fibres can be placed following the direction of the maximum stresses to the structure.

The right sequence in Figure 10.5 shows a number of plies that contain aligned fibres that can be joined to produce a multi-layer unidirectional composite structure. The left sequence of Figure 10.5 shows plies that contain aligned fibres that can be joined in different orientations to produce a quasi-isotropic composite. In this case, a 0°/±45°/90° composite is formed.

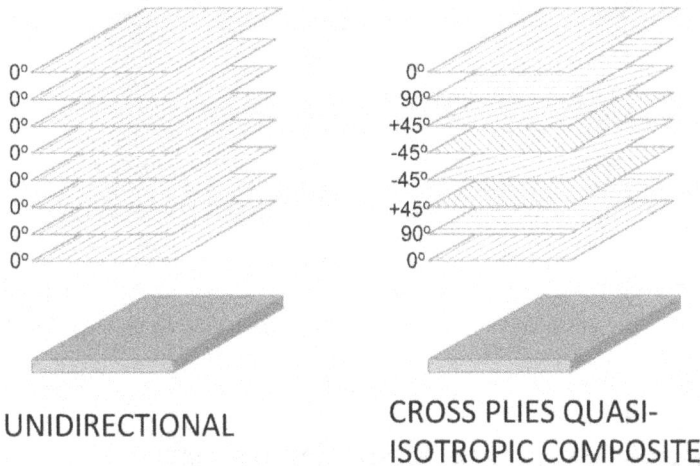

FIGURE 10.5 Different laminates created by stacking plies with different orientations.

The fibres can also be organised in a three-dimensional pattern. Even considering the simplest of weaves, the fibres on each layer of fabric have a small degree of orientation in a third direction. Further three-dimensional reinforcement can be achieved when the layers of fabrics are woven together. Figure 10.6 reflects the tensile strength variation while the fibre orientation changes in reinforced epoxy fibreglass E.

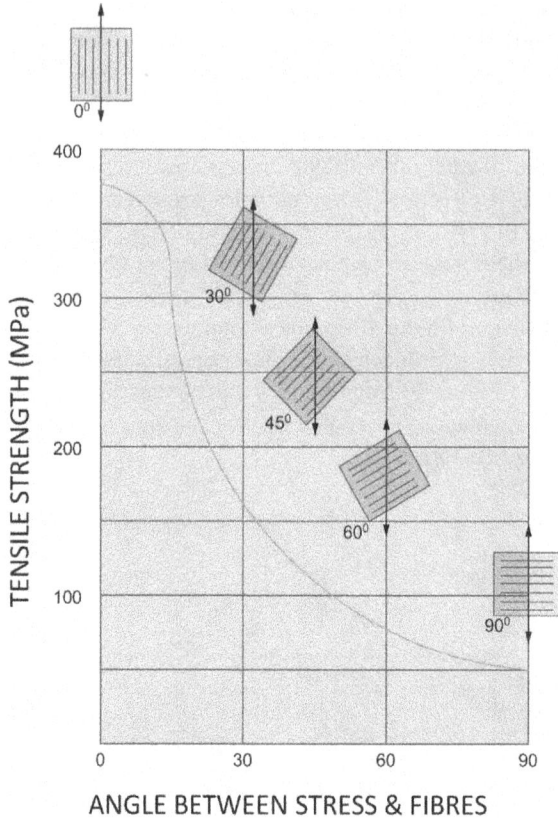

FIGURE 10.6   Representation of the effect of fibre orientation on tensile strength of reinforced epoxy fibreglass E.

10.3   QUANTITY AND PROPERTIES OF THE FIBRE

A large number of fibres increases the resistance and rigidity of the composite materials, since they possess the mechanical properties of the material.

Quantity of fibres is also known as "Fibre volume fraction" (FVF), which is the ratio of the volume of fibres to the total volume of the composite (including both fibres and the matrix). It is typically expressed as a percentage.

$$FVF = \frac{Volume\ of\ Fibres}{Total\ Volume\ of\ Composite} X^{100\%}$$

A high FVF (>60%) increases strength, stiffness, and load-carrying capacity, while a moderate FVF (30–60%) provides a balance between strength and processability. Low FVF (<30%) is easier to manufacture but has reduced mechanical properties.

On the other hand, a percentage of fibre volume greater than 80% is not recommended since it can lead to voids, and the fibres are no longer fully contained by the matrix; therefore, the transfer of loads is no longer efficient.

10.4 MATRIX PROPERTIES

The matrix properties play a crucial role in the overall performance of fibre-reinforced composites. The right matrix must complement the fibre type, ensuring effective stress transfer, environmental protection, and dimensional stability. Selecting the optimal matrix involves balancing mechanical performance, toughness, processability, and resistance to environmental factors to meet the demands of specific applications.

Fibre-matrix interaction will have a direct impact on performance, so it is important to consider these three factors:

- **Matching Modulus of Elasticity:** The matrix and fibres must have compatible stiffness to avoid differential stresses, which can lead to cracking or delamination.
- **Fibre Volume Fraction (FVF):** The matrix must sufficiently wet and encapsulate the fibres. Inadequate wetting leads to voids and poor load transfer.
- **Thermal Expansion Compatibility:** If the matrix and fibres have different coefficients of thermal expansion, thermal cycling can cause internal stresses or micro-cracks.

10.5 BOND BETWEEN FIBRES AND THE MATRIX

The bond between fibres and the matrix is essential to the strength, durability, and longevity of composite materials. It ensures efficient stress transfer, prevents failure mechanisms like delamination and fibre pull-out, and enhances the overall mechanical properties. A combination of surface treatments, proper material selection, and optimised processing conditions is necessary to achieve a strong fibre-matrix bond, ensuring that the composite performs reliably under various mechanical and environmental conditions.

10.5.1 DIFFERENT MECHANISMS OF FIBRE-MATRIX BONDING

10.5.1.1 Mechanical Bonding
- Surface roughness of the fibres allows the matrix to penetrate and physically lock into the fibre surface, enhancing adhesion. This is mainly effective with fibres that have textured surfaces (e.g., glass or carbon fibres).

10.5.1.2 Chemical Bonding
- Chemical interactions between the fibre surface and the matrix will create a covalent or ionic bonds. Also, surface treatments promote chemical bonding with specific matrices.

10.5.1.3   Interfacial Shear Strength

- Strong adhesion increases the interfacial shear strength (ISS), which prevents fibre pull-out under stress. The higher the ISS, the better the load transfer.

10.5.2   Factors Affecting Fibre-Matrix Bonding

10.5.2.1   Fibre Surface Treatment

- Coatings or treatments modify fibre surfaces to improve compatibility with the matrix.

10.5.2.2   Matrix Type and Compatibility

- Polar matrices, like epoxies, bond better with fibres that have polar functional groups. Thermoplastics may need additional coupling agents to ensure proper adhesion.

10.5.2.3   Fibre Wetting

- Proper wetting ensures that the matrix fully covers and penetrates the fibre surface. Inadequate wetting can lead to voids, reducing the interfacial bond strength.

10.5.2.4   Processing Conditions

- Curing temperature and pressure affect the quality of fibre to matrix bonding. Poor curing can lead to weak adhesion and void formation. Thermal expansion mismatches between the fibre and matrix can also cause stress and reduce bonding accuracy.

10.5.3   Impact of Poor Fibre-Matrix Bonding

- **Fibre Pull-out:** Fibres slip out from the matrix under stress, leading to premature failure. This occurs when the bond strength or interfacial shear strength is low.
- **Delamination:** Layers of the composite separate under stress, reducing strength and stiffness. This is common in laminates where interlayer adhesion failed due to poor interface.
- **Micro-cracking:** Weak bonding can lead to small cracks at the interface, which propagate under cyclic loading, causing fatigue failure.

10.5.4   Methods to Improve Fibre-Matrix Bonding

- **Optimizing Surface Roughness:** Increasing fibre surface roughness increases mechanical interlocking.
- **Choosing Compatible Matrix-Fibre Pairs:** Ensuring that the matrix and fibre materials have similar polarities and chemical properties will improve bonding between them.
- **Controlling Process Parameters:** Applying the right temperature, pressure, and curing cycle will ensure proper bonding between the matrix and the fibres.

BIBLIOGRAPHY

'Analysis of Composite Materials. A Survey'. Z. Hashin. *Journal of Applied Mechanics*, Vol. 50 (1983), No. 3, 481–505.

'Composite Materials: Design and Applications'. Daniel Gay, Suong V. Hoa and Stephen W. Tsai. CRC Press LLC, 2003.

'Composite Materials Handbook. Volume I – Polymer Matrix Composites Guidelines for Characterization of Structural Materials – MIL-HDBK-17'. McGraw Hill/Departments and Agencies of the Department of Defence, 2002.

'Engineering Mechanics of Composite Materials' (Second edition). Isaac M. Daniel and Ori Ishai. New York: Oxford University Press, 2006.

'Fiber Composite Analysis and Design: Composite Materials and Laminates. Volume I'. Z. Hashin, B. W. Rosen, E. A. Humphreys, C. Newton and S. Chaterjee. Washington, DC: U.S. Department of Transportation: Federal Aviation Administration Office of Aviation Research, 1997.

'Introduction to Composite Materials Design'. Ever J. Barbero. USA: Department of Mechanical & Aerospace Engineering – West Virginia University/Taylor & Francis, 1998.

'Marine Composites' (Second edition). Eric Green Associates. MD: Eric Green Associates Inc., 1999.

'Structural Composite Materials'. F. C. Campbell. ASM International, 2010.

11 Fabric Structures

Fabric structures in composites refer to how reinforcing fibres are arranged to enhance specific mechanical properties. These structures can vary depending on how the fibres are aligned, woven, or interlocked.

Below are the main types of fabric structures used in composite materials:

- Surface mat or fibreglass tissue
- Chopped fibre mat
- Continuous fibre mat
- Unidirectional (UD)
- Woven fabrics
- Multiaxial
- Knitted fabrics
- Braided fabrics
- Non-crimp fabrics (NCF)

11.1 SURFACE MAT OR FIBREGLASS TISSUE

It is a low-grammage C glass, normally covering a range between 25 and 80 gr/m^2. The main characteristic of this thin tissue is that it is isotropic (equal strength in all directions) and easily mouldable. Figure 11.1 shows a sample of how a surface mat or fibreglass tissue looks. It is mainly used in the naval industry to be placed in contact with the gelcoat, which is the first commonly used coat when starting a new laminate

FIGURE 11.1 Sample of how a fibreglass tissue looks.

DOI: 10.1201/9781003565222-11

sequence. Surface mat fabric is combined with gelcoat with the intention of providing greater resistance to this first coat and generating a chemical barrier against the external environment. This is possible due to its low grammage, allowing high resin absorption to gain these characteristics. On the other hand, it has lower strength compared to aligned fibres.

These fibreglass tissues are essential when complicated shapes need to be conformed, such as small radii or corners, preparing the surface for the upcoming layers.

11.2 CHOPPED FIBRE MAT

A chopped fibre mat, as shown in Figure 11.2, is a type of non-woven reinforcement fabric made from randomly distributed short fibres or filaments cut to a certain length, generally between 40 mm and 50 mm, typically bonded together by a chemical binder to form a mat. They can be easily found on the market with a wide range of grammages, and they are widely used in composite manufacturing because they offer good isotropic properties and are easy to conform to complex shapes.

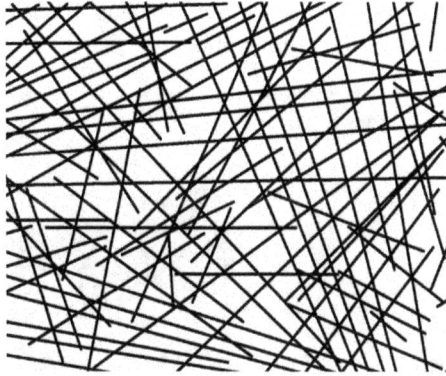

FIGURE 11.2 Representation of a chopped fibre mat configuration.

11.3 CONTINUOUS FIBRE MAT

A continuous fibre mat, as shown in Figure 11.3, is a type of reinforcement in which fibres run continuously throughout the mat, providing better mechanical properties than chopped fibre mats. It is shown in a similar way to the chopped filament mat in terms of grammage, but unlike chopped fibres, continuous fibre mats consist of long fibres that extend across the entire length of the composite structure, resulting in superior strength and stiffness.

The main difference lies in the capacity for deformation, as due to their structure it is possible to replicate complicated and deeper shapes in all directions.

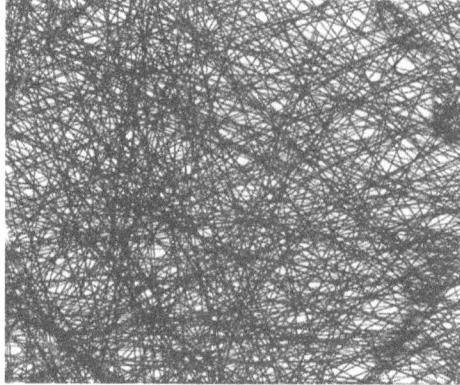

FIGURE 11.3   Representation of a continuous fibre mat configuration.

11.4   UNIDIRECTIONAL

Unidirectional (UD) fabric is a high-performance reinforcement material in which all fibres are aligned in a single direction (typically 0°). This alignment maximises the material's strength and stiffness along the fibre axis, making it ideal for applications where loads are applied in a specific direction. Figure 11.4 shows a typical 30mm width unidirectional fabric roll of 300g/m².

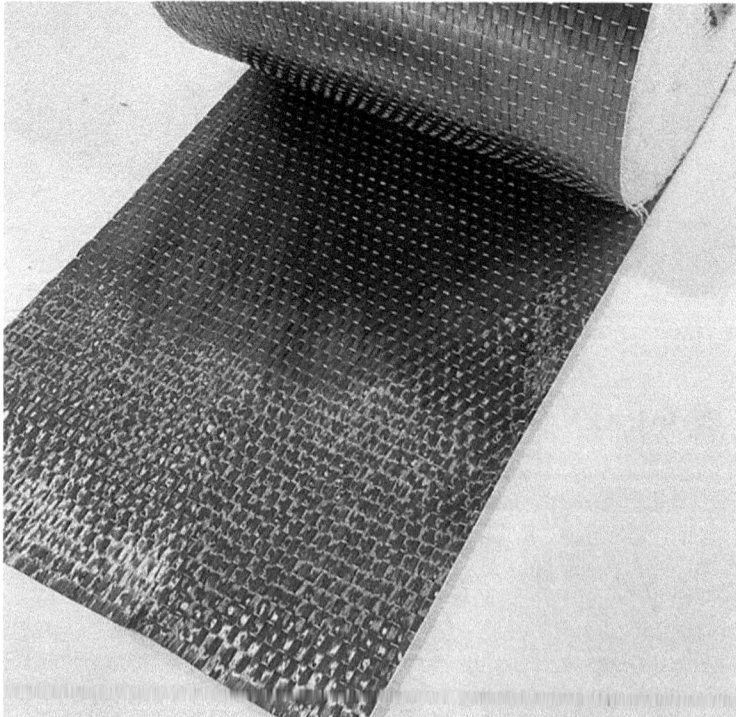

FIGURE 11.4   Carbon fibre unidirectional (UD) fabric.

11.5 WOVEN FABRICS

Woven fabrics are reinforcement structures where fibres are interlaced at right angles (warp and fill) to form a stable, textile-like fabric. Figure 11.5 shows a typical woven diagram where the warps and fills cross perpendicularly with each other. The warp filaments run along the longitudinal direction of the roll, and the cross filaments run along the fill direction.

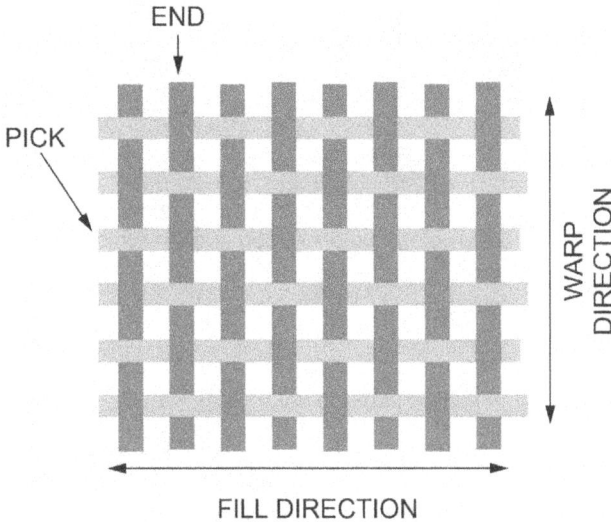

FIGURE 11.5 Typical woven diagram.

Fibres alternate between passing over and under each other, forming a specific weave pattern, where the resistant directions are generally oriented in the fill and warp directions (0° and 90°).

These fabrics are popular in composite manufacturing because they provide strength in two directions and good dimensional stability.

Woven fabrics are commonly made with carbon, glass, aramid, or natural fibres embedded in polymer matrices (such as epoxy or polyester) to create strong, light-weight composites.

11.5.1 TYPES OF WOVEN STRUCTURES

11.5.1.1 Plain Weave

Plain weave fabric is the most basic and commonly used woven structure for reinforcing composite materials. In a plain weave, warp and fill fibres are interlaced in a simple one-by-one pattern. As shown in Figure 11.6, each warp fibre alternates over and under each fill fibre, creating a tightly packed and stable structure. This pattern provides high dimensional stability, making it well-suited for applications that require strong, flat surfaces.

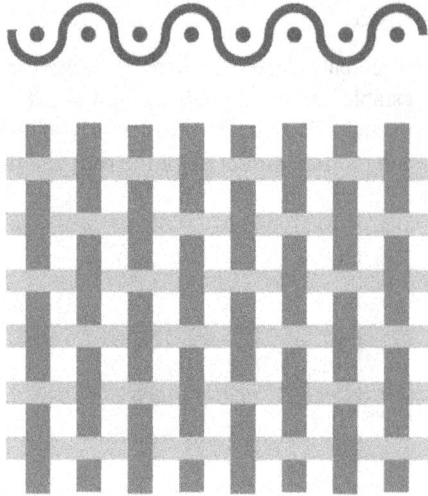

FIGURE 11.6   Plain weave diagram.

11.5.1.2   Twill Weave

Twill weave fabric is a type of woven reinforcement used in composite materials. It is characterised by its diagonal pattern, as shown in Figure 11.7, where each fibre in the warp or fill direction passes over two or more adjacent fibres, rather than alternating over and under as in plain weave. This arrangement improves drapability, flexibility, and surface finish, making twill weave ideal for applications requiring curved or complex shapes.

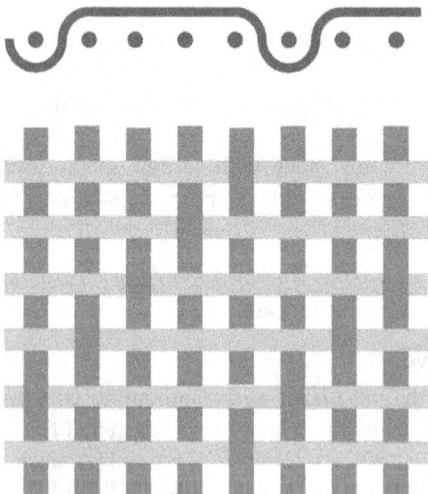

FIGURE 11.7   Twill weave diagram.

11.5.1.3 Satin Weave

Satin weave fabric is a high-performance reinforcement different from the previously mentioned types, where each warp or fill fibre filament passes over several other fibres before interlacing. Figure 11.8 shows a clear diagram for this weave configuration. The result is a smooth surface with minimal crimping, making it ideal for applications where aesthetics, high drapability, and strength are important.

FIGURE 11.8   Satin weave diagram.

Satin weaves offer superior conformability to curved or complex shapes compared to plain or twill weaves.

11.6 MULTIAXIAL

Multiaxial fabric consists of multiple layers of fibres aligned in different directions. Unlike traditional woven fabrics, where fibres are interlaced, multiaxial fabrics feature straight, non-crimped fibres stitched together in varying orientations. This structure maximises strength along multiple axes, making multiaxial fabrics ideal for applications requiring anisotropic strength and high stiffness.

11.6.1 Structure of Multiaxial Fabrics

Multiaxial fabrics are made by stacking and stitching together multiple layers of unidirectional fibres in multiple orientations. Different configurations could be found depending on the orientations of the fibres:

- Biaxial (0°/90°)
- Triaxial (0°/±45°)
- Quadriaxial (0°/90°/±45°)

Each fibre layer is aligned along a specific axis to optimise strength and stiffness along each direction. These fabrics are known as non-crimp fabrics (NCF) due to the absence of fibre interlacing, reducing fibre damage and improving load transfer.

11.7 KNITTED FABRICS

Knitted fabrics, as shown in Figure 11.9, are created by interlocking loops of yarn to form a textile structure. Unlike woven fabrics, which use a crisscross pattern in their fibres, knitted fabrics are typically made from a single continuous thread. Due to this configuration, knitted fabrics perform with greater elasticity and flexibility. In composite applications

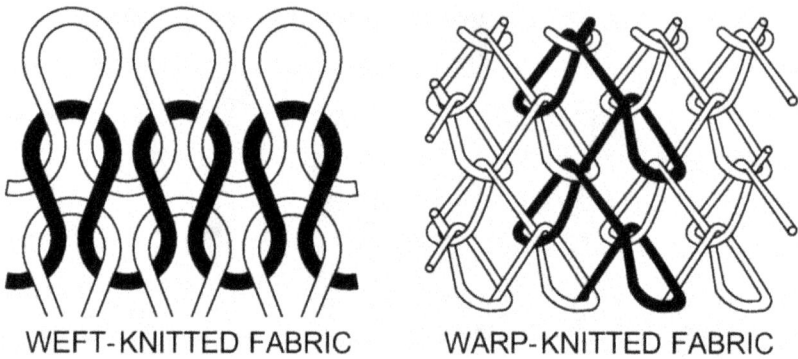

WEFT-KNITTED FABRIC   WARP-KNITTED FABRIC

FIGURE 11.9 Knitted fabric interlocking loop pattern representation.

11.8 BRAIDED FABRICS

Braided fabrics are made by interweaving multiple strands of fibres in a braiding pattern. Figure 11.10 illustrates the braiding process, where fibers are continuously interlaced around a mandrel by rotating carriers, forming a braided structure as the mandrel moves forward.

This technique creates a three-dimensional structure, resulting in fabrics that provide excellent mechanical properties and unique characteristics.

Braided fabrics are special for applications where high strength, flexibility, and impact resistance are essential.

11.9 NON-CRIMP FABRICS (NCF)

Non-crimp fabrics (NCF) are made of non-crimped fibres that, in most cases, are held together by stitching, as shown in Figure 11.11, or bonding, which helps to maintain fibres aligned and placed or oriented in a specific direction. The fibres are laid parallel to each other, minimizing fibre crimp (bending), which is common in woven fabrics.

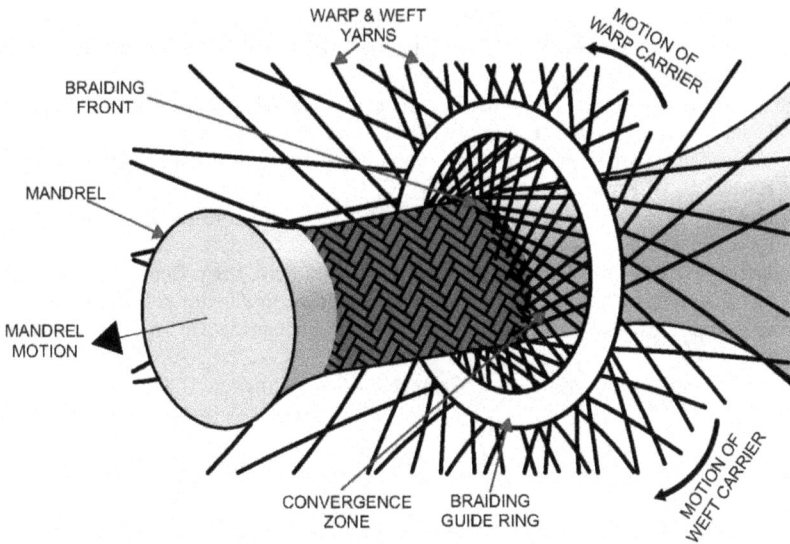

FIGURE 11.10   Braided fabric interweaving representation.

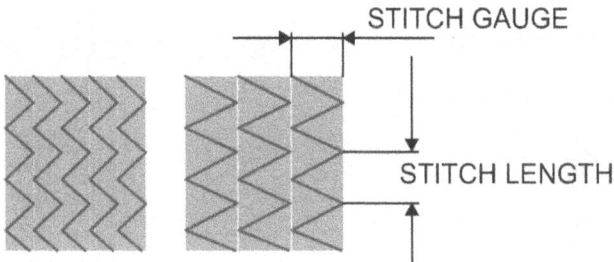

FIGURE 11.11   Non-crimp fabric representation.

The most common fibre types used in NCF are carbon fibre, glass fibre, and aramid fibre, where multiple layers of unidirectional fibres are stacked and stitched together.

BIBLIOGRAPHY

'Analysis and Performance of Fiber Composites' (Third edition). Bhagwan D. Agarwal, Lawrence J. Broutman and K. Chandrashekhara. Wiley India Pvt. Ltd., 2015.

'Analysis of Composite Materials. A Survey'. Z. Hashin. *Journal of Applied Mechanics*, Vol. 50 (1983), No. 3, 481–505.

'Composite Materials: Design and Applications'. Daniel Gay, Suong V. Hoa and Stephen W. Tsai. CRC Press LLC, 2003.

'Composite Materials Handbook. Volume I – Polymer Matrix Composites Guidelines for Characterization of Structural Materials – MIL-HDBK-17'. McGraw Hill/Departments and Agencies of the Department of Defence, 2002.

'Engineering Mechanics of Composite Materials' (Second edition). Isaac M. Daniel and Ori Ishai. New York: Oxford University Press, 2006.

'Fiber Composite Analysis and Design: Composite Materials and Laminates. Volume I'. Z. Hashin, B. W. Rosen, E. A. Humphreys, C. Newton and S. Chaterjee. Washington, DC: U.S. Department of Transportation: Federal Aviation Administration Office of Aviation Research, 1997.

'Introduction to Composite Materials Design'. Ever J. Barbero. USA: Department of Mechanical & Aerospace Engineering – West Virginia University/Taylor & Francis, 1998.

'Marine Composites' (Second edition). Eric Green Associates. MD: Eric Green Associates Inc., 1999.

'Structural Composite Materials'. F. C. Campbell. ASM International, 2010.

12 Thermoset Matrices

A thermoset polymer can be defined as a type of polymer material that cannot be reversed from its hardening process when heated or chemically cured. This means that it does not soften or flow once it has cured, no matter how much the temperature it is exposed to; it will decompose before it flows.

There is also no chemical substance that allows it to dissolve. This hardening process, called cross-linking, forms a rigid, three-dimensional network structure that gives thermoset polymers their durable, heat-resistant, and insoluble properties. For this reason, they are materials that, once cured, cannot be recast or reprocessed, unlike thermoplastics, which can be melted and reshaped multiple times.

The initial state of thermosetting matrices is a viscous liquid, which, due to the effect of a hardening reaction, will first go through a gel state until it finally becomes a solid material.

The brittleness and resistance to temperature of the final product will be defined by the cross-linking. The higher the thermal resistance, the higher the brittleness will be; in contrast, the ability to absorb energy will be lower, as will the stretching resistance and chemical resistance.

The main advantages of thermosetting matrices in the final product are high rigidity, low weight, high thermal stability, resistance to creep and deformation under loads, and good thermal and electrical insulation properties.

12.1 TYPES OF THERMOSETTING MATRIX

Common examples of thermoset polymers include epoxy resins, phenolic resins, and polyurethane. These materials are used in a wide range of applications, such as adhesives, coatings, and electrical insulation. The most commonly used resins in the naval industry are as follows:

- Polyester resins
- Vinylester resins
- Epoxy resins

12.1.1 POLYESTER RESINS

Polyester resins are synthetic resins made through the polymerization of esters. These resins are typically unsaturated, meaning they contain double bonds in their chemical structure, which allows them to cure (harden) through cross-linking when exposed to a catalyst, which is often an organic peroxide. Once cured, polyester resins form a hard, durable, and water-resistant material with excellent mechanical properties at a low cost compared to the other matrices.

DOI: 10.1201/9781003565222-12

They are the most used resins among all the thermosetting matrices and are widely used in various industrial applications due to their strength, versatility, and affordability. Due to their characteristics, they are mostly used in the construction of boats, especially in high-demand production. In general, these types of resins have a low glass transition temperature, and their resistance and rigidity are not very high. Something to remark is that during the hardening process, they tend to contract, being this one of their weak points. To put this in numbers, they contract between 6% and 10%.

The viscosity at room temperature of the polyester resins used in hand lay-up laminates is around 300 cPs (centipoise), although there are specific resins, such as those for infusion, in which the viscosity is established around 100 cPs.

Polyester resins could last for six months at room temperature; this number could be shorter if the container has been opened. Humidity, high temperature, and sunlight must be avoided to preserve the product. They also require the use of a solvent, which is styrene, and also of a cobalt accelerator in the curing process.

12.1.1.1   Polyester Resins Categorization

Polyester resins can be categorised into unsaturated polyester resins and saturated polyester resins.

- **Unsaturated Polyester Resins (UPRs):** These are the most common type of polyester resins and are widely used in applications where strength and durability are required, such as in fibreglass-reinforced plastics (FRP). These resins cure through a process called cross-linking, where a catalyst (like methyl ethyl ketone peroxide or MEKP) initiates a chemical reaction on a mixed resin to conform to the desired material.

- **Saturated Polyester Resins:** Different from the unsaturated resins, saturated polyester resins do not contain double bonds and, therefore, do not cure through cross-linking. Instead, they are used in coatings, adhesives, and paints due to their flexibility, durability, and resistance to chemicals and moisture.

12.1.1.2   Cobalt Accelerator (Cobalt Octoate)

It is the most common accelerator used in the curing of polyester and vinylester resins. It is a cobalt-based compound that is added to the resin to increase the rate of curing; this means it is used specifically to speed up the resin curing or hardening process.

When cobalt accelerator is added to a resin, it reacts with the peroxide catalyst (such as MEKP), activating the decomposition of it to form free radicals, which initiate the curing process. This reaction can happen at lower temperatures than without the accelerator, which can be useful in certain manufacturing processes.

Cobalt accelerator is typically added in small amounts, ranging from 0.5 to 3.0% by weight, depending on the specific resin formulation and curing requirements. The curing time required for the resin to harden could be reduced if the right percentage (max) of cobalt is added, or it could be increased (longer curing time) if the opposite occurs. It is important to use the correct amount of accelerator, as too much can cause the resin to cure too quickly and result in excessive heat generation, warping or cracking, making the final product not as expected.

We can identify this product as a purple liquid and it should be handled with care and the recommended equipment. It is also important to store it in a well-ventilated area, away from heat or flame.

12.1.1.3 Styrene Monomer

Styrene monomer plays a crucial role in unsaturated polyester resins; the main purpose is to copolymerize the linear chains of polyesters, acting as a cross-linking agent. Figure 12.1 shows the styrene monomer chemical structure. Styrene is also used as a reactive diluent in the production of unsaturated polyester resins for coatings and composites, to achieve final product properties such as the right viscosity, water resistance, and cost. This is essential to improve performance, facilitate processing, and ensure cost-effectiveness.

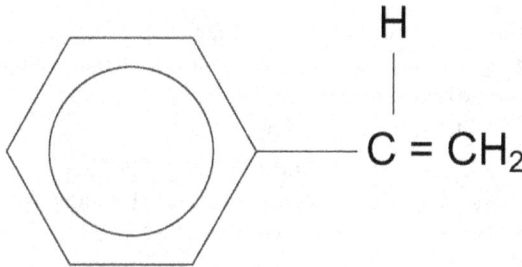

FIGURE 12.1   Styrene monomer chemical structure.

Styrene is an excellent diluent to achieve a proper processing viscosity for polyester resins.

During the curing of the resin, the formation of polystyrene polymers occurs, and a certain proportion evaporates. For all of this, the use of a slight molar excess of styrene is justified to guarantee a proper resin curing process.

There are optimum monomer concentration limits, and as a general rule, no more than 40% should be added. In case this limit is exceeded, some properties are affected; for example,

- An excess of styrene causes the resins to be brittle and sensitive to heat.
- They will have lower resistance to the environment, which means that after a few weeks or months of sun exposure, they will show superficial cracks.

Viscosity Reduction: Styrene reduces the viscosity of polyester resin, allowing better performance for applications, particularly in fibreglass-reinforced composites. This lower viscosity improves resin flow through the fibres while applying, improving the wet-out of fibres and facilitating the manufacturing process.

Cross-Linking Agent: In unsaturated polyester resins, styrene monomer actively participates in the curing (hardening) process. When an initiator like MEKP is added, styrene reacts with the polyester chains through a process called copolymerization, leading to cross-linking, which creates a rigid thermoset network that provides mechanical strength, chemical resistance, and durability to the final cured resin.

Improved Performance Characteristics: Styrene's chemical structure is in charge of forming a stable network to improve the thermal and mechanical properties of the cured resin, which directly leads to the material's strength, durability, and flexibility.

Cost Efficiency: Styrene helps to lower the overall cost of the resin while maintaining desirable properties.

12.1.1.4   Types of Polyester Resins

Depending on the type of alcohols and acids used, different types of polyester resins are obtained. According to the nature of their constituent monomers, they are divided into:

- Orthophthalic: They are the most frequently used and are the lowest cost among the polyester resins. These resins are often used in applications that do not require extreme chemical or thermal resistance, and they can absorb up to 2.5% of water if exposed for a long period of time immersed.

- Isophthalic: They have better mechanical properties than orthophthalic and better resistance when exposed to water (less absorption). They are often preferred for applications requiring higher durability and are classified as a medium to high-performance type of polyester resin. Most of the gelcoats used in the naval industry are formulated on an isophthalic base.

- Bisphenol: They have better mechanical and chemical properties than orthophthalic and isophthalic resins, although they are expensive. These resins offer exceptional chemical resistance, thermal stability, and mechanical strength, making them suitable for applications requiring extreme durability. It is the most suitable resin among polyesters for corrosive media.

12.1.2   Vinylester Resins

Vinylester resins can be defined as a midway resin between epoxy and polyester, combining the best features of polyester and epoxy resins into one solution. These resins are created by reacting epoxy resin with unsaturated carboxylic acids, resulting in a resin that combines the toughness of epoxy with the processing and curing characteristics of polyester.

They are typically stronger than polyesters and more resistant than epoxies. They have better mechanical, chemical, and thermal properties than polyester resins; they have high resilience resistance (the deformation energy that can be recovered from a deformed body when the stress ceases), good resistance to fatigue, and shrinkage during curing is less than in polyester resins (1%).

The viscosity is similar to polyester resins, which makes it easier for impregnation and manoeuvrability during moulding. They also have good adhesion to the reinforcing fibres and good fire resistance.

In the naval industry, they are mainly used for the construction of boat hulls or parts that are constantly underwater. They are also used in combinations with polyester resins in the first layers of laminate to prevent osmosis. Although it must be mentioned that these are used less frequently than polyester, as they are more expensive.

12.1.2.1   Vinylester Resin Properties

High Chemical Resistance: Vinylester resins have outstanding resistance to a wide range of chemicals, including acids, alkalis, and organic solvents. This makes them suitable for use in corrosive environments where both chemical and moisture resistance are essential.

Moisture Resistance: Compared to polyester resins, vinylester resins absorb less water, which prevents degradation over time in wet or marine environments.

Improved Mechanical Properties: Vinylesters have higher tensile strength, elongation, and impact resistance compared to polyester resins, giving them greater flexibility and reducing the risk of cracking or brittleness under stress.

Thermal Stability: These resins can withstand higher temperatures compared to polyester resins. On the other hand, they generally do not perform as well at extreme temperatures as pure epoxy resins.

Better Adhesion: Due to their epoxy base, vinylester resins bond better with reinforcing fibres like fibreglass and carbon fibre.

12.1.3   Epoxy Resins

Epoxy resins are the most widely used resins in high-quality composite materials, basically because they have better physical and mechanical properties than polyester and vinylester resins, including a very stable fluid. They have a really good adhesion ability, and as a result, the laminates will have a high number of fibre content in the final product.

It is only when mixed with an epoxy hardener that they can undergo the chemical reaction needed to cure properly; it is important to note that this is different from the cobalt compound used in polyester resins. If applied without the hardener, the resin would remain in a liquid state indefinitely.

Epoxy resins cure through a chemical reaction with a hardener (or curing agent). The curing process can be adjusted to control the resin's working time, cure rate, and final properties; it can occur at room temperature or be accelerated with heat, depending on the application requirements.

Mixing epoxy resin and hardener is an important step, from casting to coating. The ratio of the two components (component A and component B) will define the quality or performance of the final product. If the amount of either component is exceeded, it can cause issues like an unfinished cure or a weakened surface layer. It is critical to get this ratio right to ensure success in the final product. The hardness of epoxies is higher than that of polyester resin and, therefore, they can operate at higher temperatures. Most epoxy resins need the input of external heat to cure, through a curing or post-curing process.

12.1.3.1 Types of Epoxy Resins

- Two-Part Epoxy: The most common type, involving a separate resin and hardener (component A and B) mixed in specific ratios. This allows customization of cure time and properties.

- Single-Part Epoxy: Pre-mixed and generally requires heating to initiate curing. These are used where exact processing and timing control are needed.

- UV-Curing Epoxy: Special formulations that cure when exposed to UV light, used in applications requiring fast setting without heat.

12.1.3.2 Epoxy Resin Properties

- **Good mechanical properties up to 180°C:** When fully cured, epoxy resins exhibit impressive mechanical properties, including high tensile, compressive, and flexural strength. This makes them highly suitable for structural applications.

- **Good resistance to chemical attack:** Epoxy resins are highly resistant to many chemicals, including acids, alkalis, and solvents. They also resist moisture absorption, which prevents weakening in humid or wet environments.

- **Good resistance to corrosion.**

- **Low shrinkage during curing (0–1%):** Unlike some other resins, epoxies cure with minimal shrinkage, which maintains the dimensional stability of components and prevents stresses on bonded surfaces.

- **Good electrical and thermal properties:** Epoxy resins have a higher heat resistance compared to polyester or vinylester resins, making them ideal for applications that require exposure to higher temperatures.

- **Good electrical insulation:** Epoxies are excellent insulators, making them widely used in electronics for potting, encapsulating, and coating to protect sensitive components from moisture, dust, and mechanical damage.

- **High Viscosity.**

12.1.4 COMPARISON OF RESIN PROPERTIES

12.1.5 MORE THERMOSETTING RESINS

Polyester, vinylester, and epoxy resins are the most popular among the naval and other industries, especially in the production of high quality composite materials. But there are also other resins used for a huge variety of products.

TABLE 12.1
Comparison of resin properties.

Resin	Density (g/cm³)	Resistance to Traction (Mpa)	Modulus of Elasticity (Gpa)	Volumetric Contraction (%)	Resistance to Stretch (%)	Thermal Resistance (°C)	Viscosity (cps)
Polyester	1.22	60–85	4.2–4.8	7–9	2.5	60–130	350–1000
Vinylester	1.12	81	3.3–3.5	3–6	6	100–140	350–500
Epoxy	1.10	90	7	1.5	1.5	60–200	350–900

12.1.5.1 Phenolic Resins (Phenolics)

Phenolic resins are among the oldest synthetic polymers. They are derived from the reaction between phenol and formaldehyde.

They are mainly used in electrical insulators, circuit boards, adhesives, and coatings. Phenolics are also used in moulded products and in the production of wood composites, such as plywood and particleboard

Properties:

Phenolic resins present high heat resistance and flame retardancy, excellent chemical and moisture resistance, as well as good dimensional stability and electrical insulation properties.

12.1.5.2 Polyurethane Resins

Description: Polyurethanes are formed through the reaction of isocyanates with polyols. They can be either thermosetting or thermoplastic, depending on their formulation.

Properties:

- Excellent abrasion resistance and flexibility.
- High impact strength and chemical resistance.
- Good thermal stability and can be formulated for rigidity or flexibility.
- Applications: Widely used in coatings, foams, adhesives, elastomers, and casting resins. Polyurethane thermosets are commonly used in automotive parts, construction materials, and footwear.

12.1.5.3 Melamine Formaldehyde (MF)

Melamine formaldehyde is a hard, thermosetting plastic with a glossy surface, created by the reaction of melamine with formaldehyde.

It is mainly used in laminates, such as Formica, tableware, electrical insulation, adhesives, and coatings.

It presents high heat resistance, good electrical insulating properties, excellent hardness, and scratch resistance, and is resistant to chemicals and moisture.

12.1.5.4 Silicone Resins

Silicone resins are made from siloxane polymers, which include a silicon-oxygen backbone. These thermosetting resins offer excellent high-temperature stability and are used in high-temperature coatings, sealants, electrical insulators, and as a binder in paints and coatings. The most common use for silicone resins is in the electronic and automotive industries.

They are considerably flexible and resistant to UV light and present outstanding thermal and chemical resistance, even at high temperatures, good electrical insulating properties, and weatherability.

12.1.5.5 Polyimide (PI)

Polyimides are high-performance thermosetting resins known for their exceptional thermal stability and strength. They are widely used in electronics, aerospace, and automotive applications, for example, in flexible circuits, insulation, and high-temperature adhesives.

They have an extremely high heat resistance and can withstand temperatures up to 300°C or even more. They also present excellent mechanical strength and chemical resistance.

12.2 RESIN CURING PROCESS

The resin curing process is defined as the transformation of the resin from an initial liquid state to a solid state. For this to happen, the addition of an extra substance such as a catalyst, accelerator, or hardener is necessary to activate the resin curing process.

This process is divided into distinct stages, as shown in Figure 12.2: working time, hardening phase, and final curing. Each phase is essential for achieving the desired mechanical, thermal, and chemical properties of the resin.

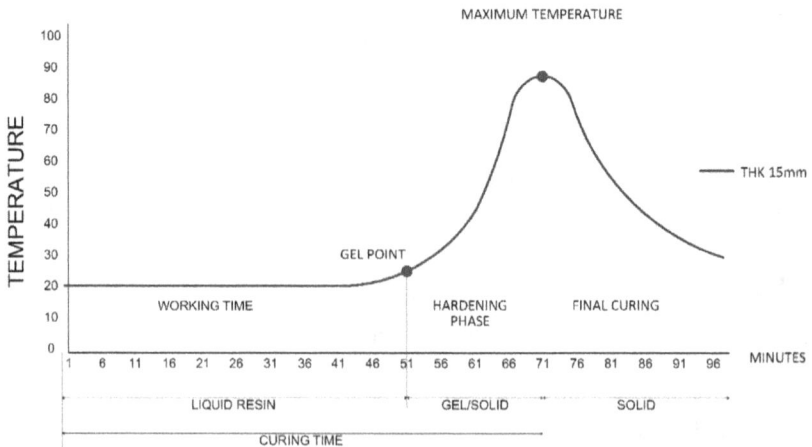

FIGURE 12.2 Typical time vs temperature diagram for a resin cure cycle.

Working Time: The working time, also known as pot life or gel time, is the period after mixing the resin and hardener. The moment the catalyst is added to the resin, the first phase of the curing process begins; this is where the resin remains workable. It is in this phase that the resin can start to be shaped into the mould or surface for laminating since the resin still does not present difficulties in flow. As time goes by, the viscosity will increase until it reaches what is called the "GEL POINT".

The gel point is the point at which the resin turns into a gel, making it impossible to continue impregnating the mould or the piece being manufactured.

There are a number of factors that can directly affect this working time, such as temperature, resin formulation, and mixing ratio. High temperatures will shorten the working time by speeding up the curing reaction, while lower temperatures will prolong it. Incorrect ratios between resin and hardener can alter the working time and affect the quality of the final cure. At last, the resin formulation will rely on the resin type and hardener and can affect how long the working time lasts.

This working time is conditioned by the percentages of catalyst in the polyester and vinylester resins, but for the epoxy resins, this is a bit more predictable, as epoxy resins with different hardeners and working times can be found, providing the option to choose the most convenient for the desired purposes.

Room temperature and the thickness of the laminate will also play an important role in the working time and curing process. Sometimes it is better to separate the laminate sequence or process into several stages to avoid overheating. Laminating many layers in one shot at a certain temperature could accelerate the hardening stage. Also, if the temperature factor cannot be avoided, it is advisable to spread the resin in different containers and mix them as needed. Preparing small volumes of mixed resin will provide more working time.

Hardening Phase: This phase, also known as gelation, marks the point where the resin mixture transitions from a liquid to a gel or partially solid state. Once the gel point has passed, the resin starts to harden, releasing heat exothermically until it reaches the maximum temperature. The resin is no longer workable, and chemical bonds are actively forming.

Temperature during this phase is crucial, as rapid heat buildup can cause distortions or defects in larger castings or thick applications.

Final Curing: Final curing, or post-curing, is the phase where the resin achieves maximum strength, durability, and resistance properties. Once the maximum temperature is reached, it begins to drop until it reaches room temperature, where the mixture becomes fully solid and the cross-linking reactions are largely complete.

There are a number of factors that can directly affect the final curing, such as temperature, humidity, mixing ratios, and the hardener (or catalyst and accelerator). Higher temperatures accelerate curing, but excessive heat can lead to thermal cracking or distortion, especially during the hardening phase. Also, incorrect ratios can lead to incomplete curing, resulting in sticky or weak final products. In terms of humidity, some resins, like polyurethanes, are sensitive to moisture, which can create bubbles or weaken the final cure. Hardeners, or additives like catalysts and accelerators, can modify the curing rate. For example, cobalt accelerators in polyester resins speed up the curing process, reducing working time and gel time.

12.2.1 FACTORS TO CONSIDER IN THE CURING PROCESS

- **Type of resin**
- **Working room temperature (ideal 17–22° C):** Be mindful of the pot life, especially in high-temperature environments or when working with fast-curing resins. Mix only as much as you can use within the working time.
- Amount of catalyst and accelerator (in the case of polyester or vinylester resins)/hardener (epoxy resins)
- **Laminate thickness:** For larger applications, work in layers, if possible, to reduce exothermic heat buildup, which can cause defects.
- **Curing or post-curing processes:** For applications requiring high strength and heat resistance, consider a post-cure heat treatment.
- **Control Temperature and Humidity:** Maintain a stable temperature and minimise exposure to humidity to ensure consistent curing.

12.3   COATING PRODUCTS

Coating products are applied to the laminate surface with the intention of achieving a better finish and, in some cases, improving aesthetics. At the same time, they protect it from environmental effects, such as chemical attacks, water, and humidity. These products are typically applied as a first layer/coat before the laminate process starts. The most popular coating used in shipbuilding is gelcoat.

12.3.1 TYPES OF COATINGS FOR COMPOSITE MATERIALS

12.3.1.1   Gelcoats

It is the first layer that is applied to the mould before starting to lay up the laminate sequence. This also provides protection against chemical attacks, weather, or humidity.

Gelcoats are polyester- or vinylester-based coatings that form a durable, glossy surface on composite products, particularly in the marine industry. They are usually isophthalic, with colour pigments and additives. Although the use of gelcoat is optional since the final colour of the product could be finalised by other methods, it should be used when we want the surface finish to meet the following characteristics:

- Colour quality and stability
- Weather resistance
- Waterproof
- Resistance to abrasion
- Prevents or avoids surface porosity
- High gloss
- Resistance to chemical attacks.

As its main disadvantage, gelcoats can be brittle and are prone to cracking under high impact or flexing, making them less suitable for highly dynamic parts.

12.3.1.2 Epoxy Coatings

Epoxy coatings are highly durable, chemically resistant, and create a strong bond between composites. They present excellent mechanical strength, high resistance to abrasion, chemicals, and moisture. Low shrinkage results in stable, protective layers.

Epoxy coatings can become yellow with UV exposure unless they are UV-stabilized or top-coated.

12.3.1.3 Polyurethane Coatings

Polyurethanes are flexible, impact-resistant, and often chosen for their UV stability, making them ideal for outdoor applications. They present UV resistance and excellent weathering characteristics, along with good chemical and abrasion resistance. They also have good flexibility, reducing the risk of cracking under impact or bending.

As a main disadvantage, polyurethanes may not adhere as strongly as epoxies and can be more susceptible to chemicals like strong solvents.

12.3.1.4 Vinylester Coatings

Vinylester coatings offer strong chemical resistance and are commonly used in composite applications requiring moisture protection. Excellent resistance to water and a variety of chemicals, including acids, are its main characteristics. They also present high toughness and impact resistance.

Due to their superior moisture barrier properties, vinylester coatings make them ideal for any marine environment, industrial tanks, pipes, and structures exposed to corrosion. But they also need extra UV protection if used outdoors for prolonged exposure.

12.3.1.5 Acrylic Coatings

Acrylic coatings are lightweight, flexible, and easy to apply, often used as a topcoat due to their aesthetic qualities. Their main properties are good UV resistance, maintaining colour and gloss in outdoor conditions, quick-drying, and easy to re-coat. They present moderate resistance to moisture and chemicals.

Frequently used on outdoor composite structures or vehicles, in aerospace for lightweight, flexible protection, and on outdoor equipment.

12.3.2 SELECTING THE RIGHT COATING

- **Environmental Exposure:** For outdoor applications with high UV exposure, such as boats or wind turbine blades, UV-resistant coatings like polyurethanes or acrylics are recommended.
- **Chemical Resistance:** Epoxy coatings are ideal for applications where chemical resistance is crucial, such as industrial tanks and pipelines.
- **Under stress exposure:** For components subject to impact or vibration, such as automotive parts, flexible coatings like polyurethane or toughened epoxy are ideal.
- **Temperature:** High-temperature applications, such as engine parts, will benefit from specialised high-temp epoxy coatings.
- **Aesthetics:** Gelcoats provide a smooth, glossy finish and are often used for cosmetic purposes on visible composite surfaces.

BIBLIOGRAPHY

'Analysis and Performance of Fiber Composites' (Third edition). Bhagwan D. Agarwal, Lawrence J. Broutman and K. Chandrashekhara. Wiley India Pvt. Ltd., 2015.

'Analysis of Composite Materials. A Survey'. Z. Hashin. *Journal of Applied Mechanics*, Vol. 50 (1983), No. 3, 481–505.

'Composite Materials: Design and Applications'. Daniel Gay, Suong V. Hoa and Stephen W. Tsai. CRC Press LLC, 2003.

'Composite Materials Handbook. Volume I – Polymer Matrix Composites Guidelines for Characterization of Structural Materials – MIL-HDBK-17'. McGraw Hill/Departments and Agencies of the Department of Defence, 2002.

'Engineering Mechanics of Composite Materials' (Second edition). Isaac M. Daniel and Ori Ishai. New York: Oxford University Press, 2006.

'Fiber Composite Analysis and Design: Composite Materials and Laminates. Volume I'. Z. Hashin, B. W. Rosen, E. A. Humphreys, C. Newton and S. Chaterjee. Washington, DC: U.S. Department of Transportation: Federal Aviation Administration Office of Aviation Research, 1997.

'Handbook of Composites' (Second edition). S. T. Peters. Mountain View, CA, USA: Process Research, 1997.

'Introduction to Composite Materials Design'. Ever J. Barbero. USA: Department of Mechanical & Aerospace Engineering – West Virginia University/Taylor & Francis, 1998.

'Marine Composites' (Second edition). Eric Green Associates. MD: Eric Green Associates Inc., 1999.

'Structural Composite Materials'. F. C. Campbell. ASM International, 2010.

13 High-Performance Thermoplastic Matrices

High-performance thermoplastics, also known as high performance polymers, are the type of materials that could provide exceptional mechanical, thermal, and chemical properties. They are designed to withstand extreme environments and stresses, which makes them suitable for applications in aerospace, automotive, medical, or even industrial fields. These types of thermoplastics have a superior resistance to chemical attacks and typically have higher melting points compared to conventional plastics. For all these reasons, they are also more expensive than conventional plastics.

Common examples of high-performance thermoplastics are as follows:

- Polyetherimide
- Polyphenylene sulfide resins
- Liquid crystal polymers
- Sulfone polymers
- High-performance polyamides
- Aromatic ketone polymers

13.1 POLYETHERIMIDE (PEI)

Polyetherimides (PEI) are a type of high-performance thermoplastic polymer known for their exceptional thermal stability, mechanical strength, and chemical resistance. They are part of a larger family of polyimide materials that present a chemical structure based on a polymer chain with repeating ether and imide groups.

One of the main properties of PEI is its thermal stability, which typically has a high glass transition temperature (Tg) around 217°C (423°F), which means they can keep their mechanical properties at elevated temperatures. It is noteworthy that they can be continuously used at temperatures up to 180°C (356°F).

PEIs also present high tensile strength and modulus, providing excellent dimensional stability and rigidity, as well as good impact resistance and toughness, making them suitable for demanding applications. They also provide resistance to a wide range of chemicals, including acids, bases, and organic solvents, making them suitable for applications exposed to potentially harsh environments.

13.1.1 APPLICATIONS OF POLYETHERIMIDES

Due to their exceptional properties, such as electrical insulating and flame retardant, PEIs are suitable for electronic components and applications where dielectric

DOI: 10.1201/9781003565222-13

strength is critical. They can also be self-extinguishing and produce less smoke during combustion. PEIs can be found in various industries, including:

- Aerospace: Components that require high-temperature resistance and low weight, such as internal parts of aircraft.
- Electronics: Used in printed circuit boards, connectors, and housings where electrical insulation and thermal stability are essential.
- Medical devices: PEIs are biocompatible and are used in devices that require sterilization, such as surgical instruments and implants.
- Automotive: Components exposed to high temperatures, such as electrical connectors and housings.
- Industrial applications: Parts that require high strength and durability, such as wear-resistant components and heat-resistant fixtures.

13.1.2 PROCESSING TECHNIQUES OF POLYETHERIMIDES

Polyetherimides can be produced using conventional thermoplastic methods, including

- Injection moulding: This is mainly used to create complex shapes and parts with small tolerances.
- Extrusion: Production of films, sheets, and profiles.
- 3D printing: Certain grades of PEI are available for additive manufacturing applications.

13.2 POLYPHENYLENE SULFIDE (PPS)

Polyphenylene sulfide (PPS) is a high-performance thermoplastic polymer known for its outstanding thermal stability, chemical resistance, and mechanical strength. PPS is a semi-crystalline polymer with an aromatic backbone and sulfur linkages, giving it a combination of rigidity, heat resistance, and chemical inertness. It is widely used in applications that demand performance in extreme environments, especially in industries like automotive, electronics, and chemical processing.

PPS's chemical resistance allows it to withstand aggressive chemicals, including solvents, acids, bases, and hydrocarbons. It is a non-flammable thermoplastic and possesses good hydrolysis resistance, maintaining its stability even when exposed to water, steam, and humid environments.

In terms of mechanical properties, PPS has high strength and rigidity, even at elevated temperatures, and excellent creep resistance, allowing PPS to withstand long-term loads.

As mentioned, PPS has a high melting point around 280°C (536°F) and its mechanical properties can also be maintained at high temperatures, often up to 200–240°C (392–464°F), which means it can resist deformation under heat, making it dimensionally stable due to its semi-crystalline nature.

PPS can be processed by injection moulding, extrusion, compression moulding, and can be reinforced with fillers to improve specific properties. It has a significantly low shrinkage, which makes it ideal for the production of parts that require high-precision tolerance.

13.2.1 Types of Polyphenylene Sulfide (PPS) Resins

PPS resins are often modified or blended with other materials to enhance specific properties:

- Unfilled PPS: Basic PPS, typically used when high chemical resistance and temperature stability are needed without reinforcement.
- Glass-filled PPS: Glass fibre reinforcement increases rigidity, strength, and dimensional stability, making it suitable for structural applications.
- Carbon-filled PPS: Carbon fibre reinforcement provides high mechanical strength, conductivity, and heat resistance, used in applications requiring high strength-to-weight ratios.
- Mineral-filled PPS: Mineral fillers improve dimensional stability and reduce cost, commonly used in applications that do not require high mechanical strength.
- Alloyed PPS: Blended with other polymers, such as polyetherimide or polytetrafluoroethylene, to create alloys with specialised properties like enhanced wear resistance or lubricity.

13.2.2 Processing Techniques of Polyetherimides

Polyetherimides can be produced by using conventional thermoplastic methods, including:

- Injection moulding: Used to produce complex shapes and high-precision parts, commonly applied for automotive and electronic components.
- Extrusion: Produces films, fibres, and rods that can be used in industrial applications.
- Compression moulding: Suitable for producing larger and more complex components.

13.3 LIQUID CRYSTAL POLYMERS (LCP)

Liquid crystal polymers (LCP) are a unique type of high-performance thermoplastics, characterised by their highly ordered molecular structure. This ordered structure gives LCPs exceptional strength, stiffness, and dimensional stability, which are maintained across a wide temperature range. These polymers are called "liquid crystals" because, like liquid crystals in displays, they form highly ordered structures that resemble those of a solid crystal, yet they retain some fluidity when melted.

LCPs are known for their excellent mechanical, thermal, and chemical resistance properties, making them ideal for applications in high-performance sectors such as electronics, automotive, aerospace, and medical devices.

13.3.1 PROPERTIES OF LIQUID CRYSTAL POLYMERS

13.3.1.1 High Thermal Stability

- High melting point: LCPs have melting points ranging from 280°C to 400°C.
- High glass transition temperature: Allows LCPs to withstand high temperatures without deforming.
- Continuous use temperature: Suitable for applications that require long-term exposure to temperatures up to 240°C, with certain grades tolerating even higher conditions.

13.3.1.2 Mechanical Strength and Rigidity

- High modulus: LCPs exhibit excellent stiffness and rigidity, which remains stable even at high temperatures.
- High tensile strength: Their molecular alignment contributes to superior tensile strength and durability.
- Dimensional stability: LCPs maintain precise dimensional stability with low shrinkage, ideal for tight-tolerance applications.

13.3.1.3 Chemical Resistance

- Excellent resistance to chemicals: LCPs resist a wide range of chemicals, including solvents, acids, and bases.
- Hydrolysis resistance: They maintain their properties in moist environments, making them suitable for humid or aqueous applications.

13.3.1.4 Electrical Properties

- Low dielectric constant: LCPs provide excellent electrical insulation, with stable properties across a wide frequency range.
- Low moisture absorption: This helps maintain electrical insulation properties even in high-humidity environments.

13.3.1.5 Flame Resistance

- LCPs are inherently flame-resistant and exhibit low smoke emission, meeting stringent fire safety standards.

13.3.1.6 Low Coefficient of Thermal Expansion (CTE)

- Dimensional Precision: LCPs have a very low CTE, comparable to that of metals and ceramics. This feature allows them to be used in applications where precise dimensional stability is essential, especially with temperature changes.

13.3.1.7 Mouldability

- Fast cycle time in moulding: LCPs can be injection-moulded with short cycle times, which increases manufacturing efficiency.
- Ability to form thin, complex parts: The flow properties of LCPs allow them to fill intricate moulds and produce very thin-walled components.

13.3.2 Types of Liquid Crystal Polymers

LCPs are often categorised based on their chemical structures, which influence specific properties. Some common types include:

- Aromatic polyester LCPs: These are the most widely used LCPs, offering excellent thermal, chemical, and mechanical stability.
- Copolyester LCPs: Modified by incorporating different monomers to achieve tailored properties like enhanced flexibility or specific processing characteristics.
- Other variants: There are modified LCPs that combine properties like enhanced wear resistance, transparency, or improved flow characteristics.

13.3.3 Processing Techniques of Liquid Crystal Polymers

Liquid crystal polymers can be processed using various thermoplastic techniques:

- Injection moulding: LCPs' unique flow characteristics allow for the creation of thin-walled and intricate parts with short cycle times.
- Extrusion: Used for making films, fibres, and profiles. The molecular alignment of LCPs in extruded products enhances tensile strength and rigidity.
- Blow moulding and compression moulding: These methods are less common but applicable for certain shapes and larger parts.

13.4 SULFONE POLYMERS

Sulfone polymers are a family of high-performance thermoplastics known for their excellent thermal stability, high strength, and transparency, combined with outstanding chemical resistance and dimensional stability. The defining feature of sulfone polymers is the presence of sulfone (SO_2) groups within the polymer backbone, which enhances their rigidity and resistance to oxidation.

Sulfone polymers are often chosen for applications that require durability under extreme environmental and thermal conditions, especially in industries like aerospace, automotive, medical, and electronics.

13.4.1 Types of Sulfone Polymers

There are several common types of sulfone polymers, each with unique properties and suitable applications:

13.4.1.1   Polysulfone (PSU)

PSU is a transparent, high-strength polymer with moderate temperature stability. It is known for its toughness, resistance to hydrolysis, and high transparency. Commonly used in medical devices, plumbing fittings, and food contact applications due to its resistance to hot water and steam sterilization.

13.4.1.2   Polyethersulfone (PES)

PES has higher thermal stability than PSU and excellent chemical resistance. It is resistant to acids, bases, and oxidation. Frequently used in food and beverage processing equipment, automotive parts, membranes, and medical devices where it may be repeatedly sterilised.

13.4.1.3   Polyphenylsulfone (PPSU)

PPSU offers the highest thermal stability and impact strength among sulfone polymers. It is highly resistant to sterilization methods, including steam autoclaving. Ideal for medical instruments, aerospace, and automotive components where durability, impact resistance, and sterilizability are essential.

13.4.1.4   Polysulfone Copolymers

Copolymers combine sulfone groups with other structural groups (e.g., ether or ketone) to tailor properties such as flexibility, processability, or cost. Used in applications requiring specific mechanical properties or more economical production processes.

13.4.2   PROPERTIES OF SULFONE POLYMERS

13.4.2.1   Thermal Stability

- High heat resistance: Sulfone polymers have a high glass transition temperature (Tg), usually in the range of 180–230°C, allowing them to maintain properties at elevated temperatures.
- Dimensional stability: Their low thermal expansion makes sulfone polymers suitable for tight-tolerance applications even under fluctuating temperatures.

13.4.2.2   Mechanical Strength

- Toughness and rigidity: Sulfone polymers exhibit excellent tensile strength and impact resistance, with PPSU being particularly known for high impact resistance.
- Creep resistance: They show low creep under long-term loading, making them reliable for structural applications.

13.4.2.3   Chemical Resistance

- Resistant to chemicals: Sulfone polymers are resistant to many aggressive chemicals, including acids, bases, oils, and certain solvents.
- Hydrolysis resistance: They resist hydrolysis in hot water and steam making them ideal for applications that require regular sterilization or exposure to moisture.

13.4.2.4 Flame Resistance

- Inherent flame retardancy: Many sulfone polymers are naturally flame-retardant and self-extinguishing, with low smoke generation, making them suitable for safety-critical applications.

13.4.2.5 Electrical Insulation

- Good dielectric properties: Sulfone polymers are excellent electrical insulators and perform well even in high-humidity environments.

13.4.2.6 Transparency

- Certain sulfone polymers, especially PSU, are naturally transparent, allowing for visual inspection of fluid flow or internal structures in components.

13.4.3 PROCESSING TECHNIQUES OF SULFONE POLYMERS

Liquid crystal polymers can be processed using various thermoplastic techniques:

- Injection moulding: LCPs' unique flow characteristics allow for the creation of thin-walled and intricate parts with short cycle times.
- Extrusion: Used for making films, fibres, and profiles. The molecular alignment of LCPs in extruded products enhances tensile strength and rigidity.
- Blow moulding and compression moulding: These methods are less common but applicable for certain shapes and larger parts.

13.5 HIGH-PERFORMANCE POLYAMIDES (HPPAs)

High-performance polyamides (HPPAs) are a category of polyamide (nylon) polymers engineered to meet the requirements of demanding applications, particularly where standard nylons (like PA6 or PA66) would fail. These polyamides are specifically designed to maintain exceptional mechanical strength, thermal stability, and chemical resistance, even at elevated temperatures or in harsh environments. HPPAs are used in various industries, including automotive, electronics, aerospace, and industrial machinery, where reliability and performance are essential.

13.5.1 TYPES OF HIGH-PERFORMANCE POLYAMIDES

Several types of high-performance polyamides are used for specialised applications:

13.5.1.1 Polyphthalamide (PPA)

PPA is a semi-aromatic polyamide that exhibits higher thermal and chemical resistance than aliphatic polyamides like PA6 and PA66. It is widely used in automotive applications for under-the-hood components, such as fuel systems, cooling systems, and transmission parts. It is also used in electronic connectors and consumer electronics.

13.5.1.2   Nylon 12 (PA12)

PA12 has lower moisture absorption and better dimensional stability compared to other nylons, along with good impact resistance. It is used in automotive fuel lines, flexible tubing, cable sheathing, and medical devices.

13.5.1.3   Nylon 46 (PA46)

PA46 has a higher melting point and rigidity than standard nylons, with excellent wear resistance. It is ideal for gears, bushings, and bearings that require high wear resistance, thermal stability, and durability.

13.5.1.4   Nylon 9T (PA9T)

PA9T is a semi-aromatic polyamide with excellent thermal, chemical, and moisture resistance. It offers outstanding dimensional stability and maintains strength at high temperatures. It is used in electronic connectors, high-temperature automotive parts, and industrial applications where precise dimensions are required.

13.5.1.5   Aromatic Polyamides (Aramids)

Aramids, like Kevlar and Nomex, are characterised by their exceptional strength-to-weight ratio, heat resistance, and flame retardancy. They are not typical thermoplastics but still fall under high-performance polyamides due to their unique properties. They are commonly used in protective clothing, aerospace components, reinforcement in composites, and safety applications.

13.5.1.6   Specialised Blends and Alloys

HPPAs are often modified with fillers or blended with other polymers to enhance specific properties, such as rigidity, conductivity, or thermal stability. These blends are used in highly customized applications across various industries where specific performance characteristics are essential.

13.5.2   PROPERTIES OF SULFONE POLYMERS

13.5.2.1   Thermal Stability

- High melting points: HPPAs have higher melting points (often above 300°C) compared to standard nylons, making them suitable for high-temperature applications.
- Heat resistance: They exhibit high glass transition temperatures (Tg) and retain structural integrity and mechanical properties even at elevated temperatures.
- Dimensional stability: These polyamides have low thermal expansion coefficients, which contribute to dimensional precision, even under fluctuating thermal conditions.

13.5.2.2   Mechanical Strength

- High tensile and flexural strength: HPPAs are strong and rigid, capable of withstanding high stresses and loads without deformation.

- Impact resistance: They show excellent impact resistance, even in low-temperature environments, making them durable in impact-prone applications.
- Creep resistance: HPPAs have outstanding creep resistance, allowing them to maintain their shape and performance under long-term loading conditions.

13.5.2.3 Chemical Resistance

- Resistance to oils, fuels, and chemicals: They are highly resistant to automotive fluids, oils, fuels, greases, and various chemicals, making them suitable for under-the-hood automotive and industrial applications.
- Hydrolysis resistance: HPPAs are resistant to hydrolysis, meaning they retain stability and mechanical properties even in humid or aqueous environments.

13.5.2.4 Friction Resistance

- HPPAs often exhibit low friction and good wear resistance, which is especially valuable in applications with moving parts or high contact points, such as gears and bearings.

13.5.2.5 Electrical Insulation

- Excellent dielectric properties: HPPAs provide stable electrical insulation, suitable for high-voltage and high-frequency applications, particularly in electronics and electrical components.
- Low moisture absorption: Certain high-performance polyamides absorb less moisture compared to standard polyamides, which helps maintain their electrical insulating properties and dimensional stability.

13.5.3 Processing Techniques of Sulfone polymers

Liquid crystal polymers can be processed using various thermoplastic techniques:

- Injection moulding: LCPs' unique flow characteristics allow for the creation of thin-walled and intricate parts with short cycle times.
- Extrusion: Used for making films, fibres, and profiles. The molecular alignment of LCPs in extruded products enhances tensile strength and rigidity.
- Blow moulding and compression moulding: These methods are less common but applicable for certain shapes and larger parts.

13.6 AROMATIC KETONE POLYMERS

Aromatic Ketone Polymers are a family of high-performance engineering thermoplastics known for their exceptional thermal stability, mechanical strength, and resistance to chemicals and radiation. The term "aromatic ketone polymers" generally refers to polymers with repeating aromatic (benzene ring) units linked by ketone (C=O) groups in the polymer backbone. These polymers, such as polyetheretherketone (PEEK), polyetherketone (PEK), and polyetherketoneketone (PEKK), are used in extremely demanding applications where most other plastics would fail.

13.6.1 Types of Aromatic Ketone Polymers

13.6.1.1 Polyetheretherketone (PEEK)

PEEK is the most widely used aromatic ketone polymer and offers an excellent balance of properties, including high temperature resistance, mechanical strength, and chemical resistance. It is used in aerospace, automotive, oil and gas, medical implants, and electronics for components like gears, seals, bearings, surgical tools, and cable insulation.

13.6.1.2 Polyetherketone (PEK)

PEK has a slightly higher melting point and thermal stability compared to PEEK. It also exhibits better wear resistance.

Applications: Primarily used in high-temperature environments, including aerospace, oil and gas, and automotive sectors where extreme heat and wear resistance are required.

13.6.1.3 Polyetherketoneketone (PEKK)

PEKK has a unique combination of rigidity, chemical resistance, and thermal stability, along with lower melting and processing temperatures compared to PEEK, making it easier to manufacture. It is common in aerospace, additive manufacturing, and defence, especially for components requiring high strength, chemical resistance, and dimensional stability.

13.6.1.4 Polyetherketoneetherketoneketone (PEKEKK)

This polymer has even greater thermal resistance than PEK and PEEK, with superior chemical and hydrolytic stability. It is a more specialised material due to its challenging processing requirements. It is used in extreme environments, such as deep-sea exploration, aerospace, and nuclear applications where ultra-high performance is critical.

13.6.2 Properties of Aromatic Ketone Polymers

13.6.2.1 Thermal Stability

- High heat resistance: Aromatic ketone polymers have glass transition temperatures (Tg) around 140–160°C and melting points often above 300°C, making them suitable for high-temperature environments.

- Dimensional stability: They exhibit low thermal expansion, allowing them to maintain precise dimensions even under wide temperature fluctuations.

13.6.2.2 Mechanical Strength

- High tensile and flexural strength: These polymers are highly rigid and capable of withstanding high mechanical loads without deforming.

- Impact resistance: They offer excellent impact resistance, even at low temperatures, and are less prone to brittle failure compared to some other high-temperature materials.

13.6.2.3 Chemical Resistance

- Resistance to chemicals and solvents: Aromatic ketone polymers are highly resistant to acids, bases, oils, fuels, and other aggressive chemicals, making them suitable for use in harsh chemical environments.
- Hydrolysis resistance: They maintain their properties in moist or aqueous environments and can withstand repeated steam sterilization, which is valuable in medical and food processing applications.

13.6.2.4 Radiation Resistance

- UV and Gamma Radiation Resistance: These polymers are highly resistant to radiation damage, making them ideal for medical devices and aerospace applications where exposure to radiation is common.

13.6.2.5 Electrical Insulation

- Excellent dielectric properties: Aromatic ketone polymers offer stable insulation properties, even at high temperatures, which is advantageous for electronic and electrical applications.

13.6.2.6 Flame Resistance

- Inherent flame retardancy: Many aromatic ketone polymers are inherently flame-resistant and have low smoke emissions, making them suitable for applications with stringent safety requirements.

BIBLIOGRAPHY

'Analysis and Performance of Fiber Composites' (Third edition). Bhagwan D. Agarwal, Lawrence J. Broutman and K. Chandrashekhara. Wiley India Pvt. Ltd., 2015.

'Composite Materials Handbook. Volume I – Polymer Matrix Composites Guidelines for Characterization of Structural Materials – MIL-HDBK-17'. McGraw Hill/Departments and Agencies of the Department of Defence, 2002.

'Handbook of Composites' (Second edition). S. T. Peters. Mountain View, CA, USA: Process Research, 1997.

'Introduction to Composite Materials Design'. Ever J. Barbero. USA: Department of Mechanical & Aerospace Engineering – West Virginia University/Taylor & Francis, 1998.

'Marine Composites' (Second edition). Eric Green Associates. MD: Eric Green Associates Inc., 1999.

14 Material Selection

Selecting the correct materials is an important step in the design process for engineering products. The performance of the desirable product is limited by the properties of the material of which it is made and by the shape or form to which this material can be formed.

However, compared to conventional engineering materials, the use of composites makes the situation a bit more complicated in terms of selection, due to composites being a combination of different materials, where we could find reinforcement, matrix, and possibly fillers and additives.

The design process starts with setting the service requirements of the component we want to manufacture. These requirements will include mechanical properties, such as

- Tensile strength
- Compressive strength
- Flexural strength
- Elongation
- Impact resistance
- Hardness
- Density

But there are also other properties or characteristics involved, such as

- Electrical properties
- Thermal properties: thermal conductivity, heat resistance, and thermal coefficient of expansion
- Chemical properties: resistance to acids, resistance to salt water and organic solvents, degree of water absorption, UV radiation, and weathering

Having dozens of fibres and resins to choose from, the composite material choices can be tough. Fortunately, the design requirements usually take place through a narrow family of materials and their properties or characteristics, which set some limitations to the decision as follows:

- Fibres are in charge of the overall mechanical properties of the structure, while the resin dictates the overall physical properties of the structure.
- The resin matrix transfers the applied forces to the combined high-strength fibres, which helps the part to prevent the formation and propagation of cracks and protects the fibres from damage due to environmental conditions such as corrosive chemicals.

DOI: 10.1201/9781003565222-14

- Stiffness or strength requirements will tell us whether glass, carbon, or aramid should be used.

- Resistance to the exposed environmental conditions (chemical, fluid, and temperature) will dictate whether polyester, epoxy, or vinylester resin should be used.

14.1 DETERMINATION OF THE CONSTITUENT MATERIAL

In the conceptual design phase, more than one material system and manufacturing method are selected to provide a wide choice of creative and innovative designs.

A commonly used method to do the first preliminary mapping of materials is the materials selection chart method, where the performance of different materials for specific engineering applications is compared.

Figure 14.1 shows a material selection chart, where all the data for each material variation is placed into different regions. Different types of charts can be plotted to help in the material selection, such as Young's modulus vs. density, strength vs. density, and Young's modulus vs. strength. In this case, Figure 14.1 is a plot of tensile strength against tensile modulus.

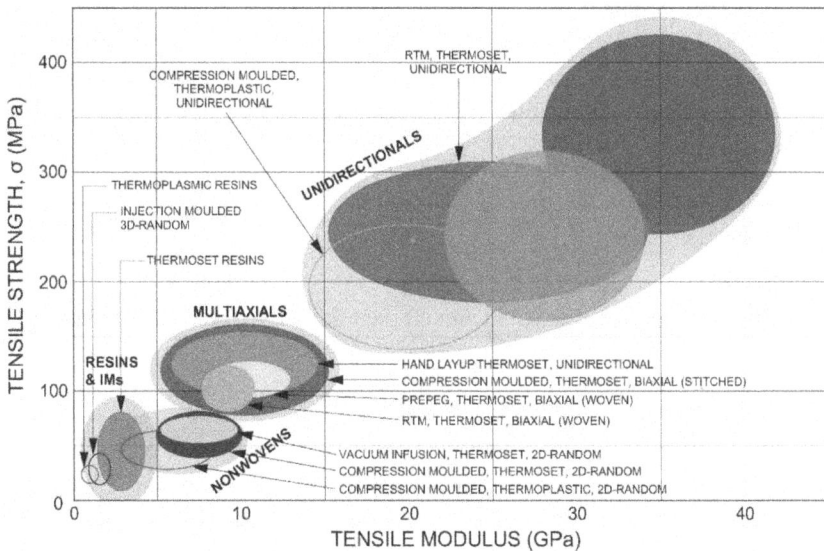

FIGURE 14.1 Materials selection chart.

Design concepts will vary, and the engineer must look at all of these options to make the right selection of material systems and manufacturing processes. Stress and other analyses must be performed to evaluate each design concept. Finite element analysis (FEA) software and other tools can greatly reduce the cost and time associated with the product development phase. A good understanding of the product requirements will help in making the right material and processing choices.

14.2 MATERIAL SELECTION METHOD

There is no standard technique used by designers to select the right material for an application. Sometimes, a material is selected based on what worked before in similar conditions or what a competitor is using in their product. The job of materials and design engineers is to consider all possible opportunities to utilise new material systems and technologies for the reduction of manufacturing cost and weight for the same or increased performance. That being said, a crucial factor that has to be considered in the material selection is the **Cost vs. Property**.

14.2.1 COST VS PROPERTY

Once a deep analysis of the product that needs to be manufactured has been performed, and knowing the loads and stresses that this future part has to support in its daily service, the path towards the choice of the correct material will be dictated by two factors:

- Cost
- Properties of the material

Composite structures, or laminates, are defined by the orientation, thickness, and material type of each ply. Multiple plies are used to build up a laminate to achieve the defined performance requirements.

- Plies with fibres parallel to the expected axial load are generally placed at a 0° orientation.
- Fibres perpendicular to the longitudinal axis will be placed at 90°.
- Fibre plies can be oriented in any off-axis direction that we want, with the main purpose of providing strength in that direction.

14.3 SPECIFIC STIFFNESS AND STRENGTH

Composite materials are ideal for applications where high strength-to-weight ratios are required. A direct indicator of this relationship is the specific resistance, which allows for comparing the structural efficiency of different materials in terms of resistance.

The plotted charts in Figure 14.2 and 14.3 show a different chart of material selection, crossing data of the specific strength against specific modulus. It can be appreciated that when the reinforcement is combined with the matrix, the specific resistance and stiffness decrease. Still, in specific applications, they are superior to metals.

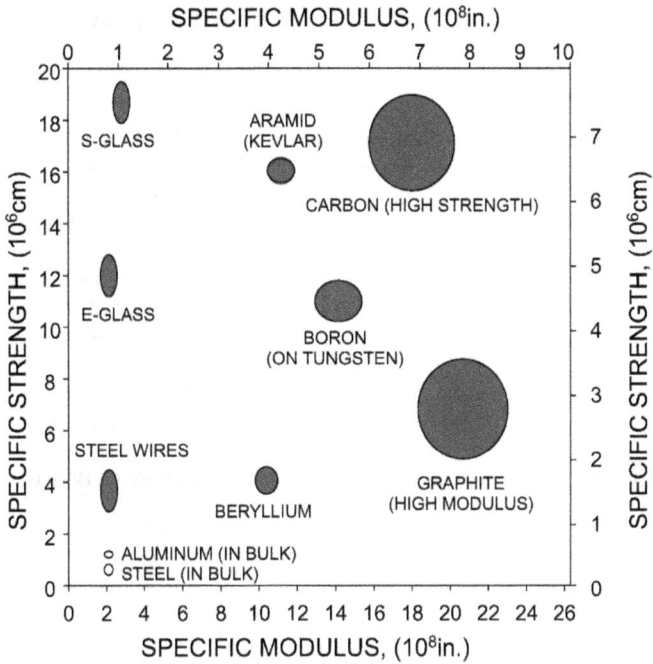

FIGURE 14.2 Specific strength vs specific modulus of different materials.

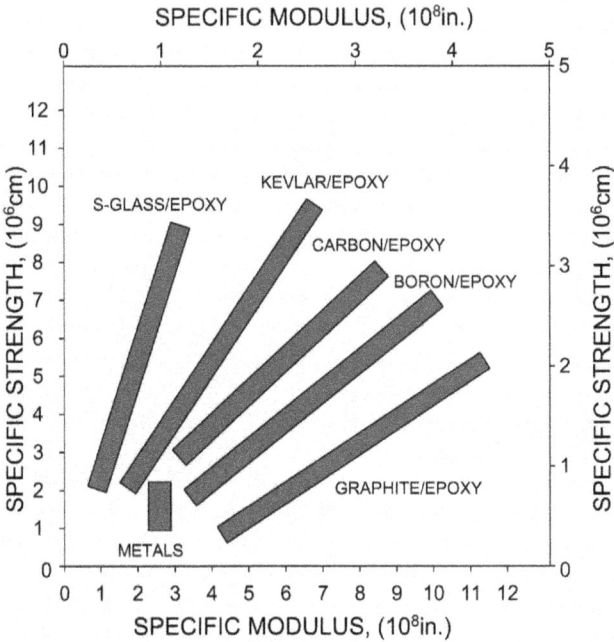

FIGURE 14.3 Specific strength vs specific modulus: Materials comparison, referring to unidirectional laminates.

BIBLIOGRAPHY

'Composite Manufacturing Process Selection Using Analytical Hierarchy Process'. A. Hambali, S. M. Sapuan, Napsiah Ismail and Yusoff Nukman. *International Journal of Mechanical and Materials Engineering (IJMME)*, Vol. 4 (2009), No. 1, 49–61.

'Composite Materials: Design and Applications'. Daniel Gay, Suong V. Hoa and Stephen W. Tsai. CRC Press LLC, 2003.

'Introduction to Composite Materials Design'. Ever J. Barbero. USA: Department of Mechanical & Aerospace Engineering – West Virginia University/Taylor & Francis, 1998.

'Mechanics of Composite Materials' (Second edition). Robert M. Jones. Taylor & Francis, 1999.

'Optimal Selection of Composite Materials in Mechanical Engineering Design'. K. L. Edwards, C. A. Abel and M. F. Ashby. Engineering Design Centre, University of Cambridge: Department of Engineering, 1994.

'Shigley's Mechanical Engineering Design' (Tenth edition). Richard G. Budynas and J. Keith Nisbett. McGraw Hill Education, 2015.

15 Manufacturing Methods

Composite materials have revolutionised industries from aerospace to automotive, as well as in the naval industry and even in the manufacture of sports equipment. The manufacturing methods for composite materials play a crucial role in their performance, cost-effectiveness, and application scope. These methods include processes like hand lay-up, vacuum bagging, filament winding, resin transfer moulding, prepreg, infusion, and automated fibre placement, each offering unique advantages for producing complex shapes, large structures, or high-performance parts.

Understanding the variety of manufacturing techniques is essential for selecting the right manufacturing method to meet specific design and application requirements. Key considerations include material properties, production volume, and structural demands, as well as factors like cost, environmental impact, and scalability. This chapter will cover a variety of composite manufacturing techniques, highlighting their operational principles, applications, and technological advancements.

15.1 HAND LAY-UP OR CONTACT MOULDING

Contact moulding is a straightforward and affordable technique used to create composite materials. In this process, reinforcing fibres are combined with liquid resin and placed onto an open mould, which may or may not have been coated previously with a layer of gelcoat for a better finish. This is one of the simplest ways to manufacture composites; contact moulding relies on resin that cures at room temperature, which keeps costs low.

Operators play a crucial role here, as they ensure the final product meets quality standards. The hand lay-up process is the first method to be used in the lamination or manufacture of composite materials. To start, the operator needs to prepare all the material to be used and decide which steps will follow to ensure the best quality possible in the final product; this includes cutting the fibre reinforcements, preparing the mould surface, coating with gelcoat if required, and calculating the amount of resin to be used.

Once all the materials are prepared and ready to start laminating, the operator will prepare the first container of resin and place the first dry layer of fibre reinforcements on top of the mould or tooling surface. Then a consolidation roller is used to spread resin onto this first dry layer, or "impregnate" it until the surface of the laminate is fully covered, to continue with the next layers and repeat the procedure. Figure 15.1 shows an operator placing a piece of fibreglass mat into the mould using the mentioned consolidation roller. While impregnating, the consolidation roller is also used to press out any air bubbles trapped between layers. It is important to note that since the pressure applied is low, the resulting laminate will have a lower fibre content, which slightly reduces its mechanical and physical strength compared to other methods. Figure 15.2 shows a representation of the hand lay-up method, including their components.

DOI: 10.1201/9781003565222-15

FIGURE 15.1 Operator placing and laminating a layer of fibreglass mat onto a mould.

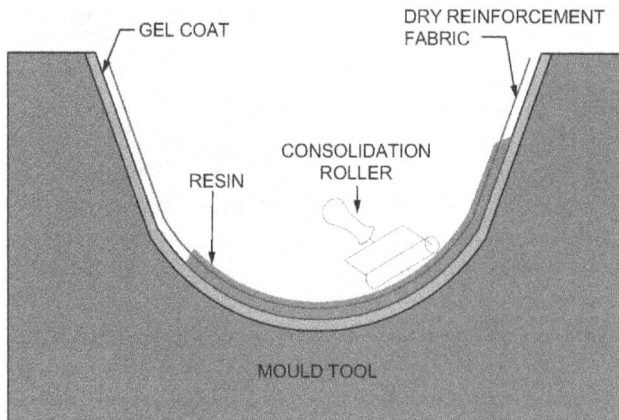

FIGURE 15.2 Representation of the hand lay-up method.

This process is the cheapest in the market due to its relative simplicity and because it does not require specialised labour compared to other methods. But it has some disadvantages as well; using this method does not ensure the best result in every work environment; the output is quite low; and it also needs a considerable amount of labour; not to mention that the ideal surface finish can be obtained on one face only (the mould face). Additionally, the final quality, as previously mentioned, depends on the operator's skills.

15.1.1 HAND LAY-UP STEPS

The hand lay-up contact moulding process is ruled by the following steps, as shown in Figure 15.3:

- Mould tool preparation and release agent application.
- Gelcoat application.

- Reinforcement preparation.
- Inspecting and preparing the mould.
- Resin preparation.
- Lamination process.
- Hardening.
- Demoulding.
- Trimming and final finishing.
- Quality control.

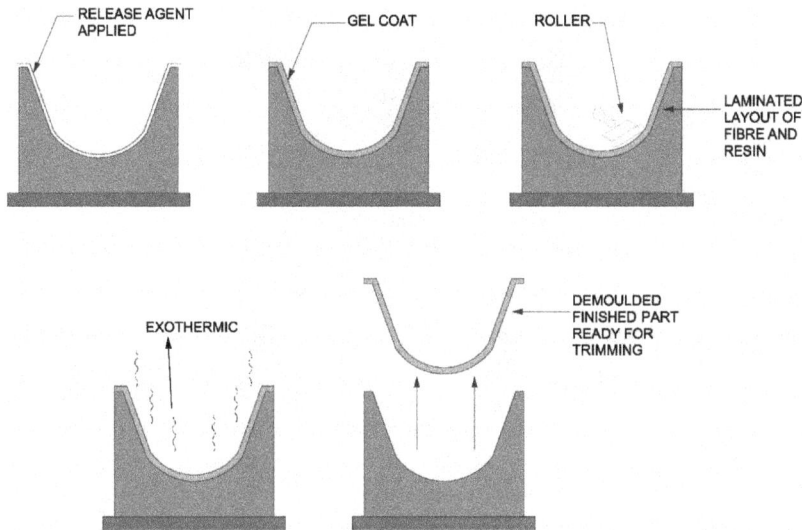

FIGURE 15.3 Schematic of hand lay-up steps.

15.1.1.1 Mould Tool Preparation and Release Agent Preparation

In the hand lay-up process for composites, mould or tooling preparation is a crucial first step that sets the foundation for a successful outcome. Preparing the mould correctly ensures that the final product will have the desired shape, surface finish, and structural integrity.

To begin, the mould surface needs to be cleaned to ensure that any dirt, dust, oils, or previous residues have been removed. This step is essential to prevent any contaminants from impacting the quality of the composite or interfering with resin adhesion, and will allow a proper demoulding process once the laminate is complete.

Once the mould surface is completely clean, a release agent is applied, typically a wax or specialised chemical, to the mould surface. This layer allows the composite to be easily separated from the mould once cured, reducing the risk of damage during demoulding and promoting the overall surface finish, Figure 15.4 represents this first step of release agent application. The release agent needs to be uniformly applied in a controlled manner, using a proper waxing and polishing technique. This waxing process will leave the surface with a shiny finish, and sometimes multiple coats need to be added for optimal results.

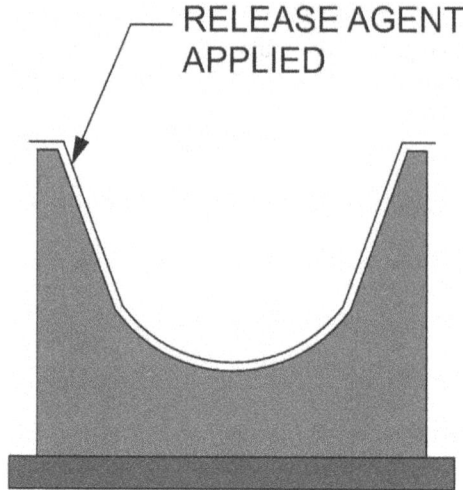

FIGURE 15.4   Hand lay-up first step: representation of release agent application.

15.1.1.2   Gelcoat Application

For applications where a high-quality and smooth surface finish is essential (due to surface roughness, chemical protection, colour, etc.), a gelcoat layer is often applied to the mould surface as a first layer. It is completely optional, but this gelcoat will act as the outermost layer of the final product and will provide a protective chemical barrier between the laminate and the external environment, as well as provide an aesthetic finish. It can also improve UV resistance, chemical resistance, and durability. Figure 15.5 shows an example of the outcome of a 40ft sail boat out of the mould with its first gelcoat layer exposed.

The gelcoat is applied and allowed to partially cure before laying down the fibre reinforcement layer and needs to be carried out as uniformly as possible: a thickness between 0.2 and 0.5 mm is standard. The gelcoat can be applied with a brush, a roller, or a spray gun, depending on the dimensions of the part and the quality required. Figure 15.6 shows a representation of an applied gelcoat layer into a female hull mould.

15.1.1.3   Reinforcement Preparation

In the hand lay-up process, preparing the reinforcement material, typically sheets of fibreglass, carbon fibre, or other fabric, is an essential step. Before starting the lamination sequence, all the reinforcement to be used on the laminate should be prepared and pre-cut according to the approximate shapes and dimensions. The type of reinforcement fabric and the fibre orientation should be selected based on the part's structural requirements, and measurements should be taken from the mould or using a template if required, to determine the exact dimensions of the reinforcement layers. This ensures that each layer will fit correctly within the mould, avoiding wrinkles or misalignments that can weaken the laminate.

Proper preparation and cutting ensure that the fibres will fit into the mould accurately, creating a strong and consistent laminate. If a combination of several

FIGURE 15.5   40ft sailboat right out of the mould with its gelcoat layer.

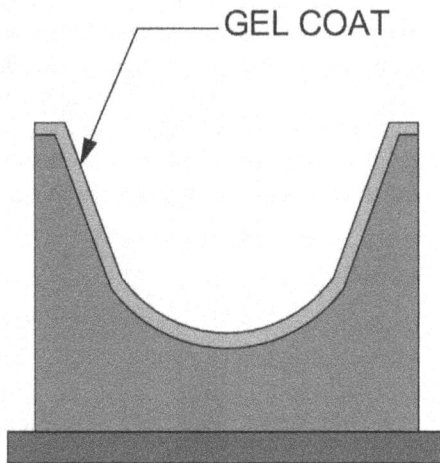

FIGURE 15.6   Representation of an applied gelcoat layer into a female hull mould.

reinforcing layers needs to be applied, all of them must be ready before starting the lamination process, since the laminate of several layers can and should be done without waiting for the previous layers to gel. Of course, this can vary

according to temperature conditions; too many layers applied at high temperatures can cause complications during the process and lead to a critical failure in
the final product.

15.1.1.4   Inspecting and Preparing the Mould

After applying the release agent and gelcoat, if used, the mould should be inspected
to ensure there are no irregularities, missed spots, or contaminants. Consistent mould
preparation will help to achieve a uniform resin flow and fibre placement, leading to
a higher-quality final product.

15.1.1.5   Resin Preparation

Resin preparation is critical to achieving a strong and durable laminate. First, the
right type of resin needs to be selected based on the part's needs. The resin to be used
during the lamination process will be prepared with the appropriate additives, such
as the accelerator (cobalt) and styrene if needed, and finally the catalyst (MEKP) or
mixed with a hardener in precise ratios to ensure proper curing immediately before
starting the lamination process. Additives, like pigments or UV stabilizers, may be
included to enhance specific properties.

To ensure good fibre saturation, resin viscosity should also be controlled, often by
warming or adding thinners if needed.

15.1.1.6   Lamination Process

For a correct and appropriate lamination process, it is necessary to start with the
application of a resin layer on top of the mould surface, or on the area where it has
been decided to start laminating, before placing the reinforcement. This will help
to settle the first layer and wet it out easily. Figure 15.7 shows a representation of
the hand lay-up lamination process into a female hull mould. Once the mould is
fully impregnated, the first reinforcing ply (previously cut according to the approximate shape and dimensions of the part) can be placed to start the laminate sequence.
A brush or roller is commonly used to apply resin to the reinforcement layer, with the
purpose of fully saturating the fibres to ensure proper bonding.

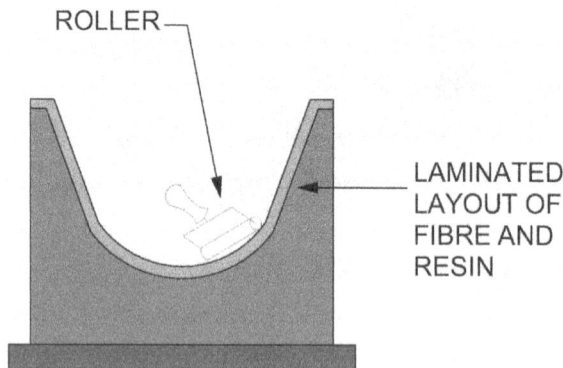

FIGURE 15.7   Representation of the hand lay-up lamination process.

It is highly recommended to start with a low grammage MAT as the first layer (fibreglass tissue); this will help to ensure the mould surface shapes correctly and also to reduce possible gaps between this first layer and the gelcoat.

Once the laminate is finished, a metallic consolidation roller is applied to remove air bubbles trapped between the layers, as shown in Figure 15.8; this also ensures resin distribution, as well as compacting the layer. It is important to take adequate time on this step, as the final quality of the product can be affected. For a thicker laminate, repeat the process by laying down more reinforcement layers, applying resin, and consolidating after each layer until the desired thickness is achieved.

FIGURE 15.8   Operator using a consolidation roller on a laminate surface.

15.1.1.7   Hardening

In the hand lay-up method, the hardening (or curing) process solidifies the resin, bonding the reinforcement layers into a rigid laminate. Once the laminated part has been fully conformed and the laminate sequence finalised, it is necessary to wait before proceeding with demoulding. It is important to allow for the initial gelation, where, after lamination, the resin begins to gel as it reacts with the hardener. This initial stage usually takes between 45 minutes and a few hours, depending on the type of resin, curing conditions, part dimensions, and rigidity of the laminate.

It is important to monitor the temperature and environment after finishing the laminate process; curing typically occurs at room temperature, but maintaining stable temperature and humidity levels helps to ensure consistent hardening. For certain resins, applying mild heat speeds up curing and enhances final strength. But do not push it; allow full cure time. The laminate should cure fully for the recommended duration (often 24–48 hours) to achieve maximum strength and durability. Figure 15.9 illustrates a composite hardening (curing) process, highlighting the exothermic reaction that occurs during the polymerization of a resin within a mould.

EXOTHERMIC

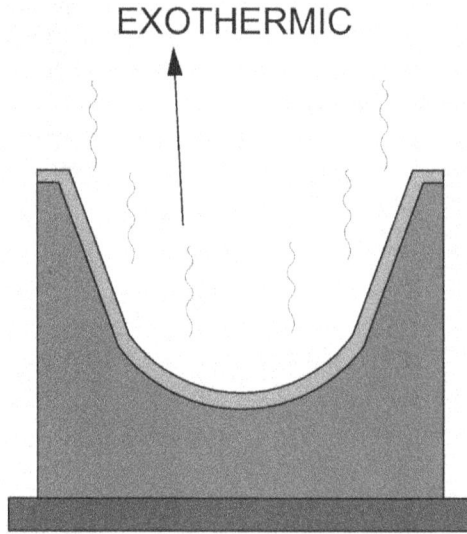

FIGURE 15.9   Representation of the hardening process.

For some high-performance applications, an additional post-cure with controlled heat is applied after initial curing to further strengthen the laminate.

15.1.1.8 Demoulding

After the lamination process is finished and the manufactured part has been fully polymerized, ensuring no deformation, the demoulding process can begin. Figure 15.10 shows a representation of composite demoulding process from a female mould. The demoulding must be performed using appropriate techniques depending on the size and rigidity of the manufactured part.

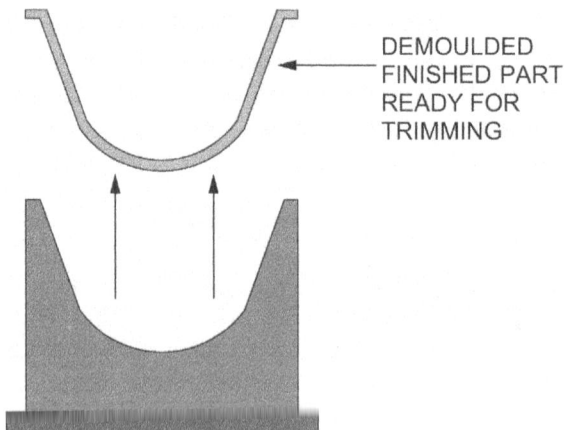

DEMOULDED
FINISHED PART
READY FOR
TRIMMING

FIGURE 15.10   Representation of the demoulding process.

To release or demould complicated parts with flat edges, the use of rigid wedges is recommended. Semi-rigid or flexible wedges are more suitable for more curved parts, since they adapt much better to the geometry without causing any damage.

For rigid moulds, it is essential to use gentle pressure or apply air to help to pop up or lift the laminate from the mould. This can be done by introducing a compressed air nozzle into a gap in specific areas of the mould. Specialized wedges or levers, designed to protect the mould and laminate, can be used for added leverage. Proper care needs to be taken to avoid using excessive force, as it could damage either the laminate or the mould.

After demoulding, inspect the laminate for any surface imperfections or defects and clean the mould to remove any residual release agent. This prepares it for the next use, ensuring consistent quality.

The demoulding process is essential for producing a defect-free laminate and preserving the quality of the mould, which can be reused in future hand lay-up processes. Proper demoulding techniques help maintain the surface finish and structural integrity of the final composite part.

15.1.1.9   Trimming and Final Finish

In the hand lay-up laminate process, the trimming and final finish steps are crucial for achieving the desired shape, size, and surface quality of the composite part.

Trimming is the removal of any excess material, such as cured resin and reinforcement fibres, around the laminate edges that extend beyond the desired part boundaries, which in hand lay-up, the excess material is quite significant. This helps the part to match the specified dimensions and makes it easier to assemble or integrate with other components.

The trimming process can be performed using proper tools like a jigsaw, band saw, or an angle grinder with appropriate cutting discs. The technicians or operators carefully cut away these excess areas; small or intricate areas may require finer tools, such as hand files or rotary tools, for precise trimming. It is essential to handle the part with care, as excessive force during trimming can create cracks or other defects.

Important note: It is very important to take safety precautions. Cutting composite materials produces dust and fine particles, which can be harmful if inhaled. Proper ventilation, dust extraction, and personal protective equipment (PPE) like respirators and safety goggles should be used to ensure safety.

At last, the final finishing will provide the desired aesthetic and protective properties for the composite part, such as gloss, UV protection, and resistance to environmental factors. Wet sanding is often used to achieve a finer finish on certain resins, especially for parts requiring a high-gloss or polished appearance, but care must be taken to avoid sanding too much, as it could expose fibres or reduce the structural integrity of the laminate. This is where attention to detail is crucial; the goal is to achieve a smooth finish without compromising the laminate layers.

For parts that need a glossy appearance, polishing compounds and buffing tools are used after sanding to achieve a smooth, reflective surface. Also, some parts may require a gelcoat or topcoat for additional protection or an improved appearance. A layer of paint, gelcoat, or even a clear protective varnish can be applied, either by brush or spray, depending on the design requirements.

15.1.1.10 Quality Control

Quality control in the hand lay-up laminate process ensures that the final composite part meets required standards for strength, durability, dimensions, and appearance. Since hand lay-up is a manual process, quality control is particularly important to maintain consistency and reliability. The finished part is thoroughly inspected for defects, including rough spots, surface bubbles, or irregularities in the finish. This is the moment to decide and apply any final touches to ensure quality.

Non-destructive tests are performed to check if the final product meets the requirements and to detect internal defects without damaging the part.

The following quality control variables have to be considered:

- Final visual appearance: surface quality, roughness, bubbles, and imperfections
- Final shape: dimensions and tolerances
- Distribution of fibre and resin
- Lack of vacuum, air, or contamination in the laminate
- Delaminated areas
- Correct final hardness

Destructive testing, such as tensile or flexural strength testing, is often conducted on sample laminates (not production parts) to verify overall quality.

Hardness testing can provide insights into the quality of curing and resin distribution.

15.2 VACUUM BAGGING HAND LAY-UP

Vacuum bagging is a technique used in conjunction with the hand lay-up lamination process to improve the quality and performance of the composite part. It could be defined as a step-up for the hand lay-up laminate process, the main difference being that pressure is applied to the laminate during the curing process using vacuum equipment.

Once the hand lay-up is finished, and there is still enough time before the polymeric matrix starts to gel, a stack of materials known as consumable materials must be placed on top of the laminate to ensure a proper vacuum bagging process. By applying pressure through vacuum bagging, excess resin is removed, and the laminate is compacted, which improves the fibre-to-resin ratio, reduces voids, and increases the strength to weight ratio of the part. Figure 15.11 shows the final hand lay-up vacuum bagging process configuration, while Figure 15.12 shows a keel box laminate vacuum bagging setup in the existing boat structure.

15.2.1 What Is Vacuum Bagging?

Vacuum is defined as the state of air pressure that is below atmospheric pressure. Therefore, the variation of pressure between the inside and outside of the bag is equal to the force applied to the laminate.

In the vacuum bagging process, pressure is applied to the laminate during the cure cycle. As mentioned, this procedure brings significant benefits to the composite

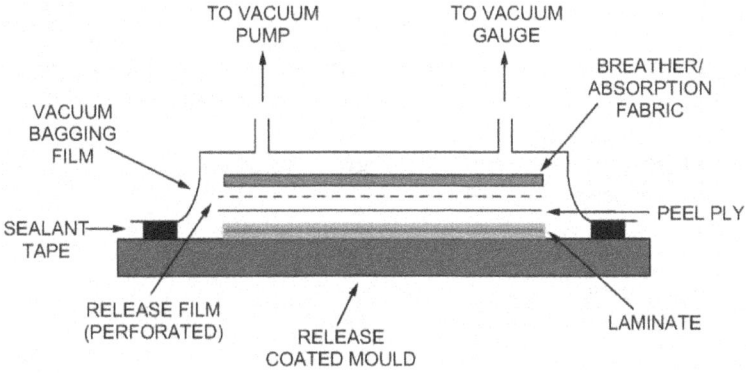

FIGURE 15.11   Vacuum bagging configuration.

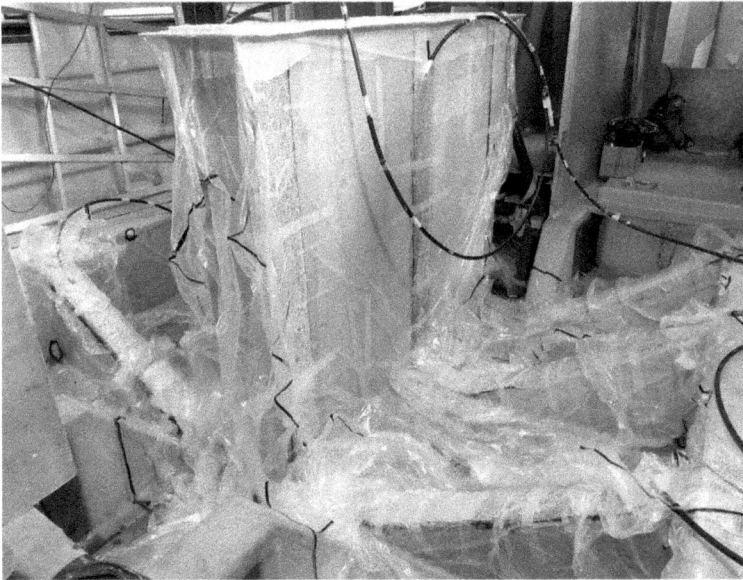

FIGURE 15.12   Keel box vacuum bagging setup on existing composite structure.

material, such as improving the strength-to-weight ratio by compacting the laminate and reducing excess resin. Vacuum bagging improves the fibre-to-resin ratio, resulting in a lighter and stronger part. This pressure also removes trapped air and minimises voids, which increases the structural integrity and surface quality of the laminate.

Because the pressure provided is uniform across the whole laminate, it creates a consistent thickness and fibre distribution throughout the part.

15.2.2 Vacuum Bagging Hand Lay-up Steps

This list of steps for the hand lay-up laminate is followed by the vacuum bagging compaction process, including the application of the consumable materials in order:

- Mould tool preparation and release agent application
- Gelcoat application
- Reinforcement preparation
- Inspecting and preparing the mould
- Resin preparation
- Lamination process
- Peel ply
- Release film or perforated plastic
- Breather
- Placing valves
- Tacky tape or sealing tape
- Vacuum bagging
- Vacuum gauge and vacuum hoses
- Apply vacuum
- Hardening
- Post curing (if required)
- Removing consumables
- Demoulding
- Trimming and final finishing
- Quality control

15.2.3 Consumable Materials Used in the Vacuum Bagging Process

In the vacuum bagging process for composite materials, various consumable materials are used to perform correct lamination, curing, and demoulding. These materials are not part of the final composite part; instead, they are removed after the curing process.

Each consumable material plays a unique role to ensure that the composite part is compacted, air is evacuated, and excess resin is controlled effectively:

- Allow compaction.
- Remove excess resin.
- Stop these materials from sticking to the part
- Facilitate demoulding
- Reduce the emissions of volatile organic elements into the environment.

a) Peel ply:

A layer of peel ply, a porous release fabric, is applied directly over the last laminate layer or ply. This material is a fabric that facilitates the extraction of the consumable materials from the laminate; it does not bond to the resin but allows resin to flow through, creating a textured surface after curing for better bonding if secondary operations are required.

Figure 15.13 shows how a peel ply fabric looks; it normally has a distinct pattern with a different colour that makes it easy to identify for removal.

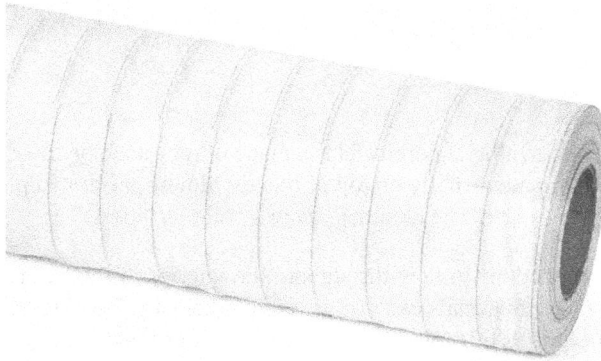

FIGURE 15.13 Peel ply fabric.

b) Release films and perforated films:

Release film and perforated plastic are used as part of the vacuum bagging stack to control resin flow, prevent sticking, and aid in part demoulding. Both materials serve to separate the laminate from the rest of the materials involved in the vacuum process, but mainly as a layer between the laminate and the breather fabric. They have specific and distinct functions and are chosen based on the specific requirements of the laminate part.

Release film is a non-porous film placed directly on the laminate to act as a barrier between the laminate surface and the breather or peel ply. Since they are resistant to resin adhesion, they prevent resin from bonding with other layers and allow for easy separation after curing, which means they can withstand the curing process. They are often made from non-stick materials like polyethylene, nylon, or fluoropolymer films, like Teflon.

The release film is used when minimal resin flow is desired or when there is no need for excess resin to escape into the breather. It provides a very smooth surface finish on the laminate but may not allow for resin reduction, as it does not let resin bleed out.

On the other hand, perforated plastics, or perforated release films, are designed to allow some resin bleed while still serving as a separation layer. The perforations enable controlled amounts of excess resin to pass through to the breather fabric, reducing excess resin and improving the fibre-to-resin ratio. Like release film, perforated plastic is typically made of materials such as polyethylene or nylon, but it contains evenly spaced holes to allow the resin to flow.

Perforated films vary by the size and spacing of the holes, allowing control over the amount of resin that bleeds out. For example, a film with larger or more closely spaced holes will allow more resin to escape. The selection of hole pattern depends on the resin viscosity, desired resin content, and lay-up thickness.

Release films are used. Perforated films are the same as release films but perforated, following a specific pattern, which allows excess resin to flow out of the laminate.

c) Breather:

This is a non-woven synthetic porous fabric, made of recycled polyester fibres that is placed over the laminate (usually on top of release film or perforated plastic) within the vacuum bagging stack. The breather has two main purposes.

- It allows trapped air to flow throughout the interior of the bag towards an exit hole. Other materials can also be used to help air flow into the bag, but the breather fulfils this role perfectly, so that the air flows inside the bag and exerts the correct vacuum pressure.
- It acts as an excess resin sucker. The excess resin flows out of the laminate due to the action of the vacuum pressure.

It is also important to use the correct amount of breather, since excess resin absorption will affect the final product, leaving the laminate with a lack of resin. On the other hand, if there is notable excess resin in the laminate, it will saturate the breather, blocking the airflow.

d) Tacky tape:

Tacky tape, also called sealant tape or vacuum bagging tape, is an adhesive material used to create an airtight seal around the vacuum bag against the mould, preventing air leaks and ensuring consistent vacuum pressure throughout the curing process. Figure 15.14 shows how a tacky tape looks.

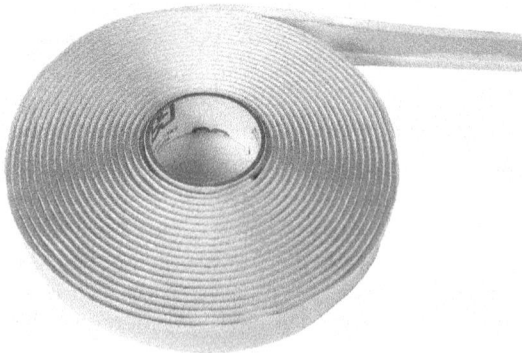

FIGURE 15.14   Tacky tape.

e) Vacuum Bag Film:

It is a flexible plastic film placed over the lay-up and sealed tightly around the edges of the mould using sealant tape, creating an airtight environment. These bags are temperature resistant and cover the entire lay-up, including the peel ply and breather fabric.

Different types of vacuum bagging films are available to withstand a range of curing temperatures. For high-temperature curing in autoclaves or ovens, heat-resistant films made of materials like nylon or polyimide are used. For room-temperature cures, less heat-resistant films such as polyethylene or polypropylene are sufficient. They can also be found in a variety of thicknesses, with thicker films offering greater durability but reduced flexibility. Thinner films are more flexible but may be more susceptible to punctures or tears.

Figure 15.15 shows a composite rudder blade that was just removed from the vacuum bag. Consumable materials can still be appreciated and are notably in order (peel ply, perforated release film, and breather).

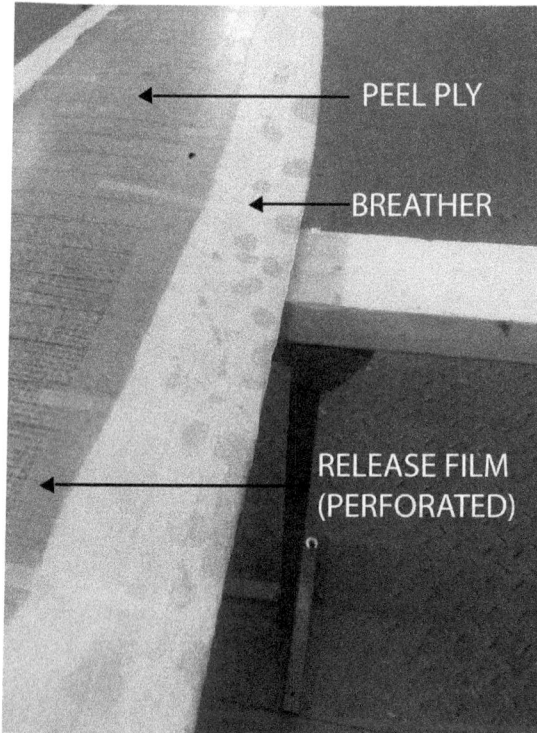

15.15   Manufactured composite rudder blade made by vacuum bagging process.

GET VACUUM?

s to get vacuum; each system is limited by its own character-
pressure that can be reached. When the bag is perfectly

sealed, the outer and inner pressure becomes equal to atmospheric pressure. The pump should achieve a vacuum around 14 psi or 711–736 mmHg (millimetres of mercury) for composite materials.

The most common systems used in shipbuilding are listed below:

15.2.4.1 Flexible Impeller Pumps
- High vacuum level
- Continuous flow
- High suction flow rates
- Electrical.

15.2.4.2 Liquid Ring Vacuum Pump
- High vacuum level
- Continuous flow
- High flow rates
- Minimum wear
- Electrical.

15.2.4.3 Vacuum Generators
- High flow
- Low vacuum levels
- Minimum maintenance
- Noisy.

15.2.5 Vacuum Bagging List of Materials

Each item plays a crucial role in ensuring the quality and structural integrity of the composite part to be manufactured. Below is a summary checklist of materials needed to perform vacuum.

15.2.5.1 Vacuum Valve and Hose
The **vacuum valve** is a special valve that connects the vacuum hose with the interior of the bag area (laminate), allowing air to be removed from within the bag. Figure 15.16 shows a common quick release valve or connector used in the vacuum bagging procedure.

A **vacuum hose** is attached on one end to the vacuum bag using the vacuum valves and to the vacuum pump on the other end. The correct diameter must be set according to the air volume that needs to be extracted.

15.2.5.2 Vacuum Pump
The vacuum pump should be capable of pulling a strong vacuum to remove air out of the bag and compact the laminate. As the vacuum is drawn, excess resin is absorbed by the breather fabric, and air trapped in the laminate layers is removed, reducing voids and bubbles.

FIGURE 15.16 Vacuum valve or connector.

15.2.5.3 Vacuum Gauge

The vacuum gauge is used to control the vacuum level inside the bag. The vacuum level is monitored to ensure consistent pressure throughout the curing process, where in some cases adjustments might be performed to maintain the desired pressure, which varies depending on the resin type and part specifications. The scale units used are bars or millimetres of mercury, as shown on the vacuum gauge in Figure 15.17. This will ensure a proper vacuum is maintained throughout the process.

FIGURE 15.17 Vacuum gauge.

15.2.5.4 Resin Traps

Resin traps or filters prevent excess resin that flows through the hoses from reaching the vacuum system. They are usually made of steel.

15.3 PRE-IMPREGNATED (PREPREG) LAMINATE

The pre-impregnated laminate (or "prepreg") process is widely used in composite materials manufacturing, especially for high-performance applications such as aerospace, automotive, and the marine industry. Prepregs consist of a reinforcement material, such as fibreglass, carbon fibre, or aramid fibre, that is pre-impregnated or coated with a thermoset, generally epoxy, or thermoplastic resin, which is partially cured or "semi-cured" (B-stage) and generally stored at low temperatures for after use when needed. This "semi-cured" stage provides the material with a semi-sticky touch called "tacking", allowing it to keep the material stable and making it easier to handle and store before final curing. Pre-impregnated materials can be found in different forms, like woven fabric sheets, unidirectional tows, and roving. Figure 15.18 shows an example of the pre-impregnation process, where the impregnation is made by rollers to allow very precise percentages of resin with uniform distribution.

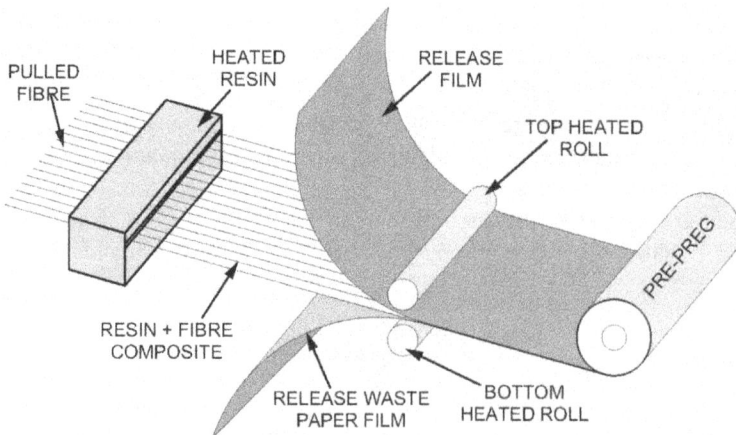

FIGURE 15.18 Prepreg impregnation process.

Prepreg materials are stored at low temperatures, often in refrigeration, to prevent premature curing. They have a limited shelf life, and all precautions must be taken to avoid any compromise in their properties.

One of the main advantages of prepregs is that the fibre/resin percentage content is perfectly controlled. In general, they tend to have a high percentage of fibre, which ensures good mechanical properties for the composite. Not to mention that the laminate process using prepreg materials is much cleaner when compared with the ones previously mentioned. There is no need to handle resins or hardeners for the mixing resin preparation, no need for brushes or squeegees, no worry about working times while trying to lay up large plies, and as a result, there are fewer man-hours and an efficient composite fabrication process.

15.3.1   LAYING-UP PREPREG

One of the most common processes of laying up prepreg laminate is placing the pre-cut layers one on top of the other. During laminate fabrication, sheets or layers of prepreg material are arranged (laying up) in a specific sequence and orientation. The orientation and layering are crucial to obtaining an optimum composite's final mechanical properties.

Generally, plastic tools with different shapes and round corners are used to help lay up the material; this will ensure there is no damage to the ply while placing it. A sharp knife is always needed to adjust, trim, or fix any situation during laminate.

15.3.1.1   Compaction and Debulking

After laying up the layers, it is highly recommended to apply a pre-vacuum called "debulking" to remove air pockets and ensure good fibre consolidation; this will ensure perfect compaction of the stack of plies. This debulking process is often achieved by applying vacuum pressure and needs to be repeated at different stages during the process until the laminate is finished.

Once the laminate process is finished, vacuum bagging is next, followed by the curing process in an oven or an autoclave.

15.3.1.2   Curing

The laid-up laminate, completely sealed and under vacuum, is then placed in an autoclave or heated press, where heat and pressure are applied to fully cure the resin. The heat allows the resin to flow, filling any gaps, and the pressure consolidates the fibres, achieving high strength and minimal voids. Curing times and temperatures depend on the resin system used. Some applications require additional heating steps (post-curing) to further enhance material properties.

15.3.1.3   Storage and Safety

Prepregs should be stored, wrapped, and sealed in polythene at a temperature around −18°C and −20°C for maximum shelf life. When the material is needed, it should be removed from the refrigeration room and must be fully defrosted before breaking the polythene seal to avoid moisture contamination. It is very important to follow the supplier's instructions in terms of storage temperature and defrosting time.

Prepregs are low risk in terms of handling hazards, but the usual precautions should be taken. Gloves and protective clothing should be worn, and mechanical exhaust ventilation should be used when heat curing prepreg systems.

15.3.2   PARTIALLY PRE-IMPREGNATED LAMINATES

Partially prepregs are comparable to prepreg materials in terms of quality but at a more reasonable cost.

The main difference is that the fibres are only partially impregnated, by selective impregnation or pre-impregnated resin films. This allows the dry fibres to act as a large, porous membrane, letting trapped air between layers escape before curing.

15.3.3 ADVANTAGES OF PREPREG

- **Consistent Resin Distribution:** Uniform distribution of resin across fibres leads to structural integrity improvement and predictable mechanical properties.
- **Control over Fibre Orientation:** Provides precise control over fibre orientation and stacking sequence, optimizing properties for specific load conditions.
- **Reduced Waste and Ease of Handling:** The partially cured resin makes prepregs easy to handle and reduces the waste associated with mixing and application of resins.

15.3.3.1 Advantages of Prepregs vs Wet Lay-up

- Low void content
- Control of fibre volume fraction
- Control of laminate thickness
- Lower labour cost
- Better quality and conformity
- Clean process.

15.3.4 PREPREG PROCESSING TECHNIQUES

- Vacuum-bag moulding
- Autoclave moulding
- Press moulding
- Pressure-bag moulding
- Thermal expansion moulding
- Tube rolling.

15.3.5 TYPES OF PREPREG

There are three different types of prepregs depending on curing temperatures:

a) **High-temperature prepregs:** They have the best physical and mechanical properties. Their curing temperature is around 180°C.
b) **Medium-temperature prepregs:** They have a curing temperature close to 120°C.
c) **Low-temperature prepregs:** They have curing temperatures ranging between 60°C and 120 °C. They are the most used in the naval industry.

15.4 PREPREG CURE CYCLE

The cure cycle in composite materials manufacturing is a carefully controlled heating and cooling process that transforms the resin within a composite part from a liquid to a solid through the application of heat. The resin within the prepreg changes from a liquid or semi-solid (B-stage) state to a fully cured, solid state. This cycle is very important in pre-impregnated composite material products since it is essential for achieving the final desired mechanical properties, and the optimal final result depends on the temperature at which the operator is working and the time it takes for laminating the part.

There are a number of stages to follow while this cure cycle is progressing. Figure 15.19 presents a generic cure cycle chart of time vs temperature, where there is an initiating ramp called "Heat-up rate" which measures how quickly the laminate and/or the mould tool reaches the cure temperature. This first phase is managed by a number of factors like matrix viscosity, matrix reaction rate, thickness of the laminate, tool mass, and tool conductivity. The next stage is the "Cure time" (or post-cure phase), followed by the "Cooling down rate" as the final stage, which controls the temperature drops to avoid thermal stresses in the component.

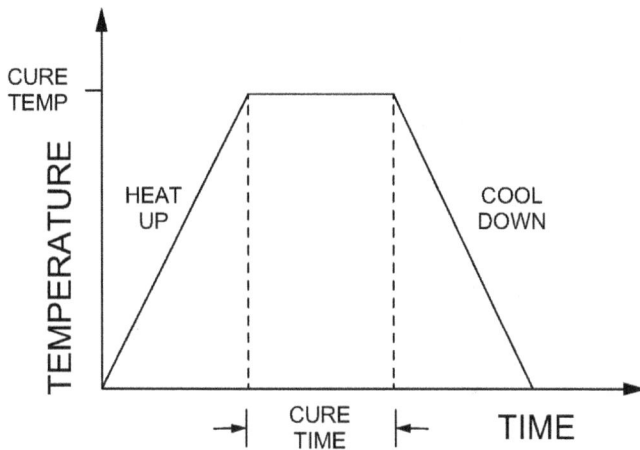

FIGURE 15.19   Generic cure cycle chart of time vs temperature.

In the "Heat-up rate", there are also sub-stages that are essential to successfully completing a cure cycle:

- Initial heat-up phase
- Dwell or soak phase
- Gelation phase

15.4.1   Cure Cycle Phases

15.4.1.1   Initial Heat-up Phase

In this stage, the composite is gradually heated at a controlled rate to avoid introducing thermal stresses. This controlled increase in temperature allows the resin to soften, facilitating flow and reducing voids. The rate of heating depends purely on the thickness of the laminate and the resin system's requirements.

15.4.1.2   Dwell or Soak Phase

At this stage, the composite laminate will be held at an intermediate temperature, allowing the resin to fully saturate the fibres and remove any trapped air. This phase also facilitates the cross-linking in thermoset resins, creating bonds between polymer chains. The dwell time and temperature depend on the specific resin system and are typically optimised to avoid excessive curing.

15.4.1.3   Gelation Phase

As the temperature continues to rise, the resin reaches the "gel point" where it changes from a viscous state to a gel. This is a critical point in the curing process, since it marks the beginning of the cross-linking. Once the resin reaches this phase, it means it can no longer flow, and its viscosity increases substantially.

15.4.1.4   Full Cure or Post-cure Phase

During this phase, the composite is held at a higher temperature (often close to or at the maximum cure temperature for the resin) for a specific period of time, allowing full cross-linking to happen. This results in a strong, rigid, and durable composite structure.

If the temperature is higher than recommended, approximately 3°C/min above optimum conditions, the resin will start to harden early, and any bubbles that may appear cannot be removed and will stay in the laminate. If the temperature is less than optimal, the resin will stay liquid for too long, and the pre-impregnated material will drain.

15.4.1.5   Cooling Down Rate

After curing, the composite is gradually cooled to room temperature, often under controlled conditions. Gradual cooling helps prevent thermal stresses or warping, particularly for thick or complex parts.

Each phase of the cure cycle is critical in controlling the properties of the final composite. Key parameters include:

- **Temperature Ramps:** Controlled temperature increases and decreases ensure even curing and prevent thermal stresses or gradients.
- **Pressure Application:** Pressure (often applied in autoclaves) consolidates the laminate, reducing voids and enhancing bonding.
- **Time at Each Stage:** The soak and cure times at each stage allow for complete chemical reactions without over-curing, which could lead to brittleness.

Important Note: For each prepreg resin system, there is a range of options for cure temperature/duration. There is also a minimum cure temperature, and for a given cure temperature, there will be a corresponding cure time. Different resin systems (epoxy, polyester, thermoplastics) require tailored cure cycles, with variations in temperature and time parameters. For example:

- Thermoset Resins: Typically require an elevated temperature cure cycle to complete cross-linking.

- Thermoplastic Resins: Generally, melt and then re-solidify upon cooling; they may not require the same curing times as thermosets.
- Other Processes: Some processes, like vacuum-assisted resin transfer moulding (VARTM) or resin infusion, may not use autoclaves and instead use ovens for curing, potentially altering the cycle parameters.

The oven or autoclave, laminate, and tooling should all reach and maintain this cure temperature throughout the specified cure cycle. Thermocouples are normally used to monitor the temperature of the laminate and the tooling while they are curing in the oven or autoclave.

15.5   LIQUID TRANSFER MOULDING (LTM)

Liquid transfer moulding (LTM) is a group of processes used in composite manufacturing where a liquid resin is injected into a mould that contains dry fibre reinforcements inside. It is important to note that this system needs to be fully sealed with the help of vacuum techniques to allow the resin to drive inside the mould and prevent any resin leakage. This means that the resin and the reinforcing fibres will start to combine for the first time inside the closed mould.

Different LTM variations can be found, like resin transfer moulding (RTM), vacuum-assisted resin transfer moulding (VARTM), and resin infusion, each designed to produce high-quality composite parts by efficiently saturating the fibres with resin. LTM is widely used in aerospace, automotive, marine, and wind energy industries because it produces strong, lightweight, and complex parts with excellent surface finishes and controlled fibre-to-resin ratios.

As with the other processes previously mentioned, liquid transfer moulding will also require steps and preparation to be performed.

15.5.1   LIQUID TRANSFER MOULDING STEPS

1- **Preparation of the Mould and Reinforcement:** The mould preparation will include surface cleaning and release agent application; some of them might require the application of heat. The pre-cut dry fibre reinforcement, which could be composed of multiple layers, is then carefully placed into the mould. Layers may be stacked in specific orientations depending on the required mechanical properties, and inserts or other components can be placed in the mould at this stage.

2- **Mould Closure:** In traditional RTM, a rigid mould or tooling is used, which is made of two or more matched metal or composite parts, which can be simplified into upper and lower moulds. These moulds are closed between each other with the help of vacuum techniques, and the resin is injected into these fully sealed assembled moulds containing the dry fibre reinforcements inside. For vacuum-assisted processes like VARTM, a flexible vacuum bag is used instead of a rigid mould top.

3- **Resin Injection:** A liquid resin, often a thermoset resin like epoxy or polyester, is injected under controlled conditions. Depending on the process:

4- **Resin Flow and Wetting:** The resin is allowed to flow throughout the reinforcement fibres until they are fully saturated. Resin flow is closely monitored to ensure that it reaches all areas, especially in complex parts.

5- **Curing:** Once the fibres are saturated, the part needs to be cured, typically at elevated temperatures. The curing process solidifies the resin and locks the fibres in place, giving the composite its strength and durability. Some LTM processes are compatible with room-temperature curing, while others benefit from higher temperatures to improve cross-linking and mechanical properties.

6- **Demoulding and Finishing:** After curing, the mould is opened, and the part is carefully removed. Minimal finishing is usually required, although trimming or post-curing (if additional heat treatment is necessary) may be performed. The final composite part generally has a smooth surface finish and high dimensional accuracy.

15.5.2 ADVANTAGES OF LIQUID TRANSFER MOULDING

- **Cost-Effective:** LTM requires fewer materials and equipment than traditional prepreg and autoclave processes, especially for large or complex parts.
- **High Quality:** With controlled resin distribution, LTM produces high-quality parts with consistent fibre-to-resin ratios and minimal voids.
- **Design Flexibility:** Suitable for producing parts with intricate shapes, large dimensions, and varying thicknesses.
- **Environmental Benefits:** Compared to open mould processes like hand lay-up, LTM reduces volatile organic compound emissions by containing the resin injection.

15.5.3 DISADVANTAGES OF LIQUID TRANSFER MOULDING

- **Process Control:** Achieving complete resin flow and wetting in complex parts requires a methodical design of the mould and resin injection system.
- **Tooling Costs:** RTM, in particular, requires high-quality, matched metal moulds, which can be expensive for low production volumes.
- **Skill Requirements:** Successful implementation of LTM demands skilled technicians with excellent attention to detail and process monitoring to prevent defects.

15.6 RESIN TRANSFER MOULDING (RTM)

RTM is a process in which the composite part is formed into a closed two-part mould (top and bottom), which could also be split into more than two parts, and resin is injected into it. This takes place by carefully placing all the dry fibre reinforcements into the open mould (lower mould). Once all the fibres have been placed and the stack of plies is complete and in order, the next step is to close the laminate using an upper mould that matches perfectly onto the lower mould, allowing a very precise gap between surfaces to be filled with the laminate. This is possible due to prior calculation of fibre thickness and resin content.

In Figure 15.20, it can be seen how a low-viscosity resin is injected or transferred fully catalysed inside the mould. The resin is injected under pressure, saturating the fibre reinforcement inside, and while it flows through the laminate, impregnating every part of it, it will also start to remove air bubbles that remain in the laminate and the mould. All the air bubbles will start to appear together with the resin through a bleed vent port that is normally on the opposite side of the mould/tooling.

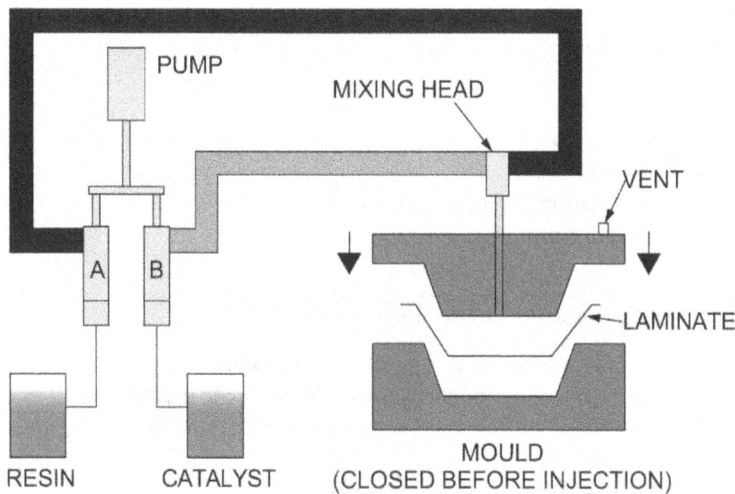

FIGURE 15.20 Representation of resin transfer moulding setup.

Once the laminate is fully impregnated (no dry areas) and the injected resin starts to show up through the bleed vent port (ensuring no more air bubbles), then it is time to interrupt or stop the resin transfer and begin the curing process.

RTM production moulds have certain requirements during the design process. The hydrostatic pressures in the closed moulds are between 2 and 10 bars, and the resin injection pressures are between 1.5 and 10 bars.

For a large production series, the moulds are usually metallic; for small productions, they can be made from composite materials. Once the moulds are closed, resin limits or boundaries must be secured. To do this, the moulds must have a proper sealing system:

a) Elastomers sealed
b) Clamping ring sealed
c) Silicone sealed
d) Resin sealed

The resin injection is run by a pumping system. The equipment is comprised of a dosing system, mixers, tanks, and controllers that ensure precise resin injection.

The resins used in RTM must have:

- Low viscosities (between 80 and 250 cPs at 20°C).
- Low exotherm to avoid problems during the process, such as contractions, deformations, and scratches on the surface.
- A long gel state transition to guarantee complete mould filling so the polymerization is not compromised.

15.6.1   RTM EQUIPMENT

- Resin injection machine
- Resin tanks
- Suitable moulds
- Temperature regulators
- Vacuum system
- Appropriate equipment for cutting dry reinforcements or pre-forms
- Polymerization cabin
- Appropriate equipment to manipulate the moulds
- Machining and trimming tools

15.6.2   MOULD TOOLING DESIGN PROCESS FOR RTM

An important detail to consider during the mould tooling design process is setting the resin entry and the bleed vent port points. The air in the mould must be evacuated to avoid future problems such as surface porosity or the reduction of the mechanical properties of the laminate.

There are three ways to introduce resin into the mould:

1) Centred injection: The resin injection point is located in the centre of the mould, and the air is evacuated from the perimeter or edges. This is the slowest filling process but needs the least amount of resin.

2) Side injection: The resin injection starts from the side and flows towards the other side of the mould, through which the air is evacuated.

3) Peripheral injection: The resin is injected all the way around the perimeter of the mould, propagating the resin towards the interior, and the air drainage is through the centre of the mould. This is the fastest filling process but also requires the greatest amount of resin.

15.6.3   TYPES OF RTM

There are three different types of RTM depending on the way the resin is transferred to the mould:

1) Standard RTM
2) Vacuum-assisted RTM (VARTM)
3) Light RTM (LRTM)

15.6.3.1 Standard RTM

A pre-shaped fibre preform is placed inside a rigid two-part mould. Designed for high-demand production, they are generally metal moulds with temperature controls that help to manage or reduce the curing times of the manufacturing product, and they are also capable of withstanding very high hydraulic pressures (from 4 to 10 bars). The resin injection pressure in this procedure is high, between 2 and 10 bars.

15.6.3.2 Vacuum-Assisted RTM (VARTM)

In VARTM, a vacuum is applied to draw resin into the mould cavity containing the fibre preform. This technique typically uses a single-sided mould with a flexible vacuum bag on the other side, making it more adaptable and cost-effective. This is a similar process to the standard RTM, but with lower injection pressures (between 1.5 to 3 bars). The moulds or tooling materials are generally made of the same composite materials as the injection pressures are lower.

Some advantages include lower tooling costs, lower resin emissions, less need for pressurization equipment, and better quality control in resin infusion. On the other hand, the surface finish is generally better on one side, as the other side uses a vacuum bag instead of a rigid mould.

15.6.3.3 Light RTM (LRTM)

LRTM is similar to traditional RTM but uses a semi-rigid upper mould and a lower vacuum bag. The mould assembly is sealed, and a slight vacuum is applied while resin is injected. This process is a lower-cost alternative to traditional RTM, and a good surface finish can be achieved. Figure 15.21 shows a 2 mm carbon fibre plate made by LRTM.

FIGURE 15.21   2-mm carbon fibre plate (C-plate) made by LRTM.

The resin transfer can be managed using a low-pressure pump (up to 3 bars) or by vacuum effect (similar to the infusion method). This method is the cheapest of the liquid transfer moulding processes, but it will have low fibre volume fractions and potentially lower mechanical properties than traditional RTM.

15.7 INFUSION

The infusion process, also called VIP (vacuum infusion process), is a technique that uses vacuum pressure to drive resin into the fibres of a laminate.

The main advantages of VIP include higher quality, better consistency, higher specific strength and stiffness, good interior finish, faster cycle time, and lower cost.

For this process, the mould or tooling needs to be fully cleaned, and the proper release agent applied. Then all the dry fibres or reinforcements are placed in the mould, whether they are MATs, fabrics, reinforcements, etc. One or several reinforcing layers can be used, and there can even be a core between layers.

This method of production is similar to the RTM, with the difference that there is no need for an upper mould to close the system. This method is assisted by the vacuum-bagging process and uses the referred consumable materials, with some slight changes. The laminate is vacuum-bagged closed with the intention of applying pressure on the laminate and generating a constant airflow with the help of a "resin flow medium", which is basically a flow mesh. Figure 15.22 is a representation of the infusion process setup applied to a hull, where there is a resin feed line located at the centreline and pumped suction lines running around the top freeboard flange. The pressurised system will drive the resin through the flow media in the direction of the vacuum lines, meaning the impregnation in this representation will be from the centreline to the sides/top of the hull, impregnating the dry fibres with the injected resin.

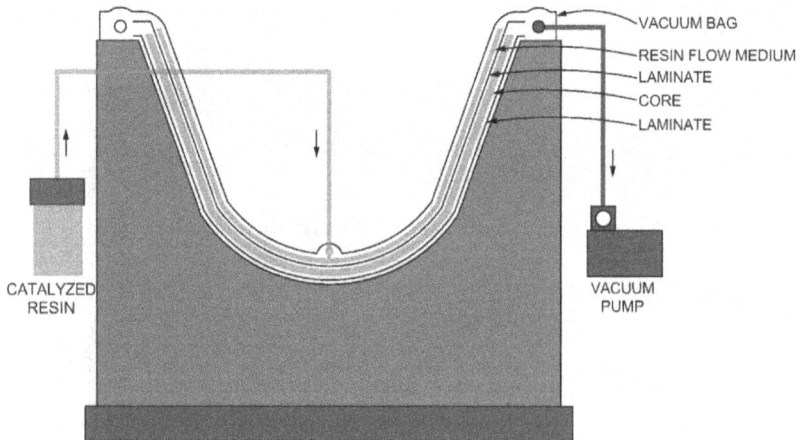

FIGURE 15.22   Representation of resin infusion setup on a hull.

15.7.1 Resin Infusion Process Procedure

Once the rigid mould is clean and prepared with a release agent to prevent the composite from sticking, and fibre reinforcements are laid into the mould following the desired orientation and thickness, the reinforcement is covered with a layer of peel ply (for surface finish) as in the vacuum-bagging process. Before jumping to a flow media mesh, a release film needs to be applied between the peel ply and the mentioned flow media mesh. This is not strictly necessary, but it can help while pulling out consumables after finishing the process. After that, a flow media mesh, which will facilitate the resin flow through the fibres, must be placed accurately. It is important to select the right flow media mesh based on the resin and the laminate sequence to be infused.

The vacuum distribution tubes (spiral wrap) covered with peel ply need to be placed peripherally, and in some cases, running on top of the laminate or through the centre line. If this is the case, care should be taken to isolate the spiral tubes from the laminate with the help of the mesh, to avoid deforming the laminate or having resin accumulation.

The next step is to place the resin feeder tubes in the previously decided strategic positions using connections such as omega flow or "T" connectors. This "omega" and the feeder tubes will not necessarily be connected to only one single feed line. For bigger projects such as hulls or decks, there will be multiple feed lines with multiple resin tanks or containers connected, so the resin flow and injection can be managed and controlled once each surface area is fully impregnated. This means that bigger projects will be built or infused in stages.

Once all the settings are complete, the full production part will be hermetically sealed using a vacuum bag, and sealant tape is applied around the edges to create an airtight seal between the mould and the vacuum bag. After a perfect seal is achieved, a high vacuum pump is used to remove all of the air in the cavity and consolidate the fibre and core materials (750 mm Hg).

Important note: It is important to ensure that everything is on schedule, that there are no vacuum leaks, and that there are no missing components. Once the resin gets in, there is no opportunity for changes during the process.

Having checked every detail, it is time to prepare the resin. The amount of resin is calculated based on the weight of the fibre and the surface to impregnate; average numbers start with a laminate with 50% fibre and 50% resin.

Still under vacuum, resin is infused into the mould cavity through strategically placed inlet ports to wet out the fabric fibres and core, ensuring complete saturation. This can be controlled by open/closed valves that allow the resin to run through. The suction is produced by the air pressure differential between areas. This makes the resin enter quickly and with a constant flow guided by the flow media, which improves distribution and minimises dry spots, perfectly impregnating all the fibres and producing a laminate without bubbles, voids, or trapped air. The speed at which the resin flows will be different for each project and will be affected by the resin viscosity and the type of fabric, fibres, etc.

Once the first area has completed the resin impregnation (considering this infusion example is made in stages), it is time to close the related feeding tube for that location, always before the resin runs out of that tank/container and the air starts to enter (this must be avoided at all costs).

After closing, for example, "line 1", "line 2" is ready to be opened, and so on, until the product is fully impregnated. With a part fully impregnated and fibres fully saturated, the vacuum continues until the resin gels. Then, and only then, can the vacuum pump be turned off.

The polymerization or curing of the resin could be carried out at room temperature or in an oven, although curing times can be reduced by adding external heat. With this technique, fibre-matrix percentages of 60–40 percent can be obtained.

The vacuum infusion process is simple in concept. However, it requires detailed planning and process design so the parts can be infused in a reasonable time without any dry spots. The rate of infusion depends on the viscosity of the resin, the distance the resin has to flow, the permeability of the media, and the amount of vacuum. Therefore, the choice of materials, flow media, resin flow layout, and location of vacuum ports are critical in making good products.

15.7.2 FACTORS TO CONSIDER IN RESIN INFUSION SELECTION

a) **Viscosity:** The lower the viscosity, the faster and more easily the resin will flow. Very low viscosities can allow air into the laminate, and there is also the risk of dry spots against the mould due to the lack of time for the resin to complete the impregnation through the laminate layers. If the viscosity is very high, the filling times can be longer and may also generate dry areas of resin due to insufficient impregnation.

b) **Gel time:** This is directly dependent on the length of the part to be laminated. Depending on the calculated infusion times, there must be a longer or shorter gel time, the only provision being that all fibres have to be fully impregnated.

c) **Reactivity:** Good resin polymerization provides laminates with better quality; otherwise, defects will appear.

15.7.3 CONSUMABLE MATERIALS USED IN THE RESIN INFUSION PROCESS

- Flow media mesh: Used as a resin flow distribution over the surface of the laminate.

- Peel ply: It also works as a distribution mesh. It helps the media mesh to distribute the resin over the surface of the laminate. At the same time, it removes air in the laminate. Once the laminate is fully cured, the peel ply allows the extraction of any excess resin remaining on the laminate together with all the consumable materials.

- Vacuum bag: Needs a good elongation property to adapt to complicated geometries. It seals the laminate and keeps the vacuum in the system to allow a successful infusion.

- Tacky tape: Seals the vacuum bag and the mould. It is important that the tacky tape has no gaps or openings; any air entering during the infusion process will affect the final result.

- Vacuum pump.

- Vacuum feed lines (resin inlet/vacuum outlet): Vinyl tubing is recommended as it is strong enough to resist collapse under pressure during the vacuum cycle.

- Resin trap: Resin traps are placed in-line to catch excess resin and protect the vacuum pump. For larger projects, multiple resin traps are recommended.

- Spiral tubing: Spiral tubing is a resin infusion standard, ideally suited to resin feed lines.

15.7.4 Ways to Introduce / Inject the Resin into the System

a) **Centred injection:** In this method, the resin is introduced from a central inlet point in the mould, allowing it to flow outward in all directions. The vacuum port is typically located around the perimeter to pull the resin toward the edges.

b) **Side injection:** Here, the resin is introduced from one side (or multiple sides) of the mould, with vacuum ports located on the opposite side. The resin travels laterally across the part, saturating the fibres as it progresses.

c) **Peripheral injection:** Resin is introduced around the perimeter of the mould, flowing inward toward a central vacuum port. The resin is infused through a series of inlets distributed along the mould's edges or through a single peripheral feed line.

15.7.5 Selecting the Correct Method

The choice of resin introduction method will depend on:

- Part Geometry: Symmetrical parts benefit from centred injection, while long or irregular parts may need side or peripheral injection.
- Part Size: Larger parts often require peripheral or multiple inlet points for complete saturation.
- Flow Path Length: Longer flow paths may need multiple injection points to prevent premature resin curing or incomplete coverage.
- Resin and Fibre Properties: Resin viscosity and fibre permeability directly affect the flow pattern and should be considered during inlet placement.

15.7.6 Advantages and Benefits of Resin Infusion

- Higher fibre-to-resin ratio (up to 70 percent fibre by weight)
- Greater strength and stiffness
- No air in the laminate/very low voids
- Very consistent laminate with good process control (fewer human errors)
- In some cases, could produce good outside and inside surface finishes
- Cleaner process
- Faster cycle time
- It is a clean process, minimizing resin waste and reducing operator exposure to volatile organic compounds.

15.7.7 Disadvantages of Resin Infusion

- Complicated setup and a need to develop the optimal vacuum ports and resin injection locations.
- In the case of a vacuum leak, the product could be spoiled.
- Higher tooling cost.

- VIP materials cost more than standard resins and fabrics.
- Expensive process, considering the number of consumable materials that are discarded.

15.8   AUTOCLAVE MOULDING

Autoclave moulding is a process in which laminated pieces, normally pre-impregnated, are exposed to higher pressure for better compaction of layers and a higher curing temperature. The process occurs in a controlled environment to ensure optimal curing and minimal defects, producing parts with excellent mechanical properties, precise tolerances, and uniform quality.

An autoclave is basically a cylindrical chamber made of heavy-duty steel or high-strength alloys to withstand high pressures and temperatures. This cylindrical vessel has to be insulated to minimise heat loss and ensure uniform temperature distribution. Figure 15.23 shows a typical autoclave chamber with a 1.6-metre diameter and 8-metre length. It can also be appreciated for having a sliding tray where the parts can be placed, vacuum connections coming from the inside, and a robust hinged door (which could also be a sliding mechanism) that has a locking mechanism to secure it.

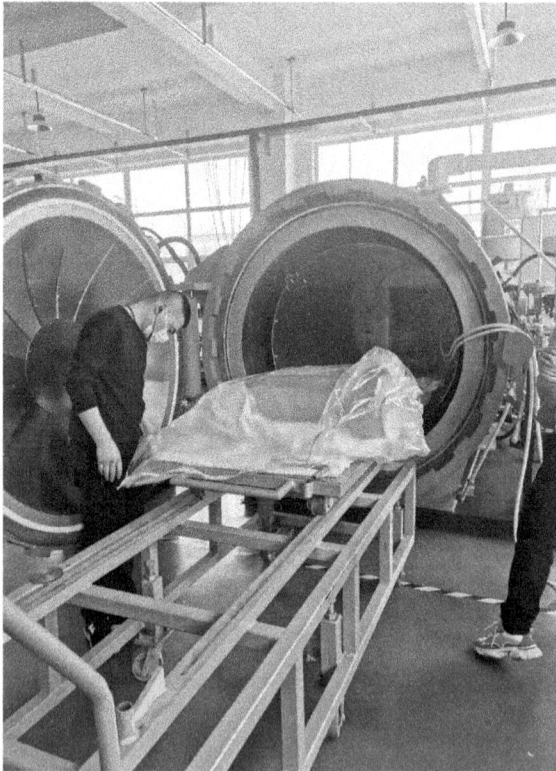

FIGURE 15.23   Standard autoclave chamber.

15.8.1 Main Components and Features of an Autoclave

15.8.1.1 Pressurized Chamber

This usually has a cylindrical surface, and the dimensions depend entirely on the dimensions and type of parts to be pressurized.

15.8.1.2 Door and Seal

Designed to maintain a leak-proof seal under high pressure. The seal used is often a high-temperature-resistant silicone or rubber gasket to ensure airtight sealing.

15.8.1.3 Heating System

Some autoclaves use built-in electric heaters to raise the temperature inside the autoclave; many others use external boilers to inject steam for heat. But the most widely used system is gas combustion, as it is cheaper than electric heating and allows temperatures up to 540°C.

Temperature controllers: Autoclaves are also equipped with sensors and controllers to maintain precise curing temperatures (120°C–180°C for composite materials).

15.8.1.4 Pressure System

Compressed air or steam is applied through an internal system that is in charge of applying pressure during the process, often in the range of 3–7 bar (45–100 psi) for composites. Gas pressurization could also be used, such as air, nitrogen, and carbon dioxide.

It is important to include a safety valve in the system to serve as a pressure relief to prevent over-pressurization.

15.8.1.5 Vacuum System

The autoclave moulding process has a vacuum system, allowing extra pressure on the manufactured part. It removes air from the vacuum bagging system or chamber to eliminate voids in the composite material.

15.8.1.6 Control System

These are devices located inside the pressurization chamber to control the process.

- **Programmable Logic Controllers (PLC):** Automates the process using a defined curing cycles by the user (temperature, pressure, and time).
- **Monitoring and Alarms:** This device provides real-time feedback on temperature, pressure, and vacuum conditions. An alarm should also be installed for any deviations from set parameters.

15.8.1.7 Cooling System

Gradually reduces the temperature after curing to prevent material stress or deformation. This is normally obtained by forced air circulation or water-cooled jackets.

15.8.1.8 Insulation and Inner Lining

Thermal insulation will ensure energy efficiency and uniform heating, while an inner lining, often made from corrosion-resistant stainless steel, will provide protection against wear and moisture.

15.8.1.9 Auxiliary Systems

- **Loading Trolleys or Racks:** They support the moulds to help get them into the chamber; this is for easy placement and removal of parts.

- **Sensors and Data Loggers:** Monitor and record process parameters for quality control.

Figure 15.24 shows a representation of the autoclave curing cycle diagram

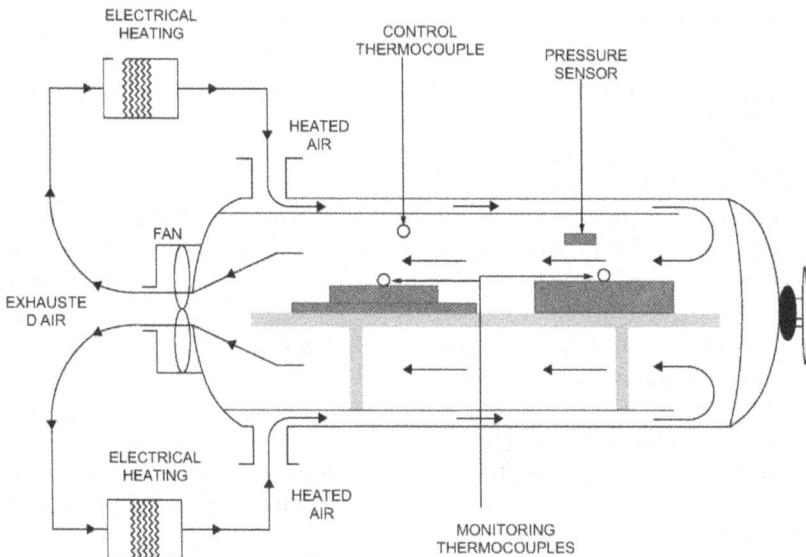

FIGURE 15.24 Autoclave curing cycle diagram.

15.8.2 Autoclave Operational Steps

- Preloading: Place materials in the chamber on moulds or trays.
- Vacuum Setup: Attach vacuum bags and ensure proper sealing.
- Process Initiation: Close the door and start the programmed cycle.
- Heating and Pressurization: The autoclave heats and pressurizes the chamber.
- Cooling: Gradual cooling ensures part integrity.
- Unloading: Remove cured parts after the cycle completes.

15.8.3 Autoclave Curing Cycle Stages

1- In the first stage, vacuum and pressure are applied for a very short time without temperature. After the desired pressure is reached, the temperature is activated and gradually increases according to the thermal ramp of the material. During this stage, the resin viscosity is reduced by the heat, allowing the resin to flow, helping air and volatile elements to escape from the laminate, as well as helping the impregnation of the fibres.

2- In the second stage, the temperature increases again until it reaches the final curing temperature and is maintained until the curing reaction is complete. To achieve a high-quality final product, both pressure and homogeneous temperature distribution inside the autoclave are essential.

The temperature distribution must be homogeneous during the entire curing process, both at the maximum curing temperature and during the curing ramp.

The results from the autoclave moulding procedure are superior to those using other methods, as are the physical and mechanical properties. The disadvantage of this procedure is its high cost, which makes it difficult to expand in the market.

15.8.4 Advantages and Disadvantages of the Autoclave Moulding Process

* Advantages
* High-quality finishes with minimal voids.
* Superior strength-to-weight ratio.
* Precise control over mechanical properties.
* Suitable for complex shapes and designs.
* Disadvantages
* High equipment and operational costs.
* Time-consuming process.
* Limited part size based on autoclave dimensions.

15.8.5 Difference between an Oven and an Autoclave

There are two main ways to cure prepreg composite materials: using a curing oven or an autoclave. The chief difference between these two is pressure. An autoclave is a pressure chamber (like the air receiver on an air compressor), while a curing oven only has atmospheric pressure.

15.9 FILAMENT WINDING

Filament winding is a machine-automated manufacturing process used to create strong, lightweight composite structures such as pipes, pressure vessels, and cylindrical or spherical parts.

As shown in Figure 15.25, this technique is performed by a machine that collects fibre rovings, pulling them from reels located and secured at a certain distance. These

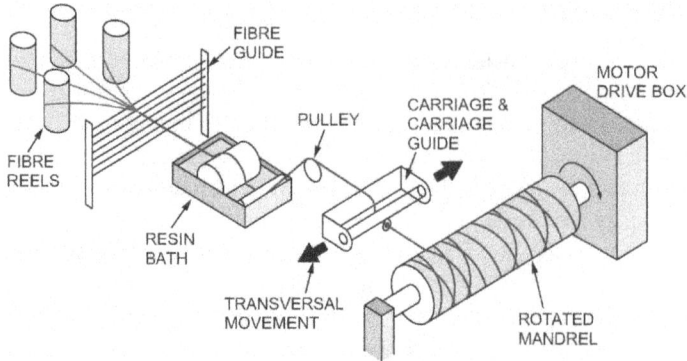

FIGURE 15.25 Filament winding diagram.

continuous fibres are then pre-impregnated or coated to be later wrapped around a rotating mandrel that has the internal shape of the desired product. Note that these fibre rovings are pulled through a resin bath immediately before being wound in a helical pattern onto the mandrel. The operation is repeated to form additional layers, each having a criss-cross pattern with the previous layer until the desired product thickness, strength, and stiffness are reached. As with most other high-quality composite processes, the resin is then cured in an oven (following the specific cure cycle), and the mandrel is removed after cooling.

The winding angle can vary from low "longitudinal" angles to high "circumferential" angles, relative to the mandrel axis. The main factor to be defined in the filament-winding process is the speed of the mandrel and the carriage. These two movements define the laminating angle, which then defines the final properties of the product. These settings can be made using a software program linked to the machine.

15.9.1 Types of Winding

Generally, there are three types of winding, as shown in Figure 15.26:

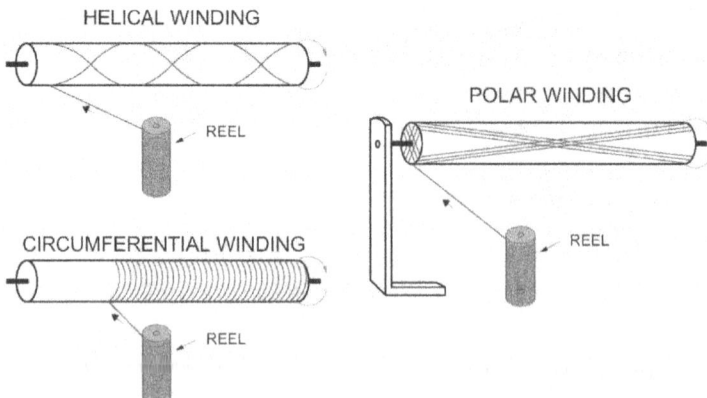

FIGURE 15.26 Types of winding.

15.9.1.1 Circumferential or Radial

The winding is performed perpendicular to the mandrel axis, providing resistance in the circumferential direction.

15.9.1.2 Helical or Axial

The fibres are placed at a certain angle in the axial direction of the mandrel.

15.9.1.3 Polar Winding

This is a special helical winding, whereby fibres run tangent to both opposite mandrel edges, creating a symmetrical plane through the laminated part. This type of winding is mainly used in spherical geometries.

As mentioned before, in filament winding, the fibres are impregnated in a resin bath or tray next to the mandrel just before they are wound. For the winding process, a wide range of resins can be used (epoxy, polyester, vinylester, phenolic), but the fibres can only be presented in reels.

It can be defined as a very fast process where the resin concentration can be controlled very accurately, reaching very high values of fibre content. One of the main benefits of filament winding is that the final cost is reduced by not using fabrics, and the structural properties of the laminates can be very high since the fibres can be placed at well-defined angles. However, the process is restricted to convex surfaces.

The structures that can be manufactured by filament winding are necessarily cylindrical, spherical, conical symmetry, or any other shape that does not have concave areas. Figure 15.27 shows a combined filament winding patterns applied onto a cylindrical surface while the mandrel is rotating.

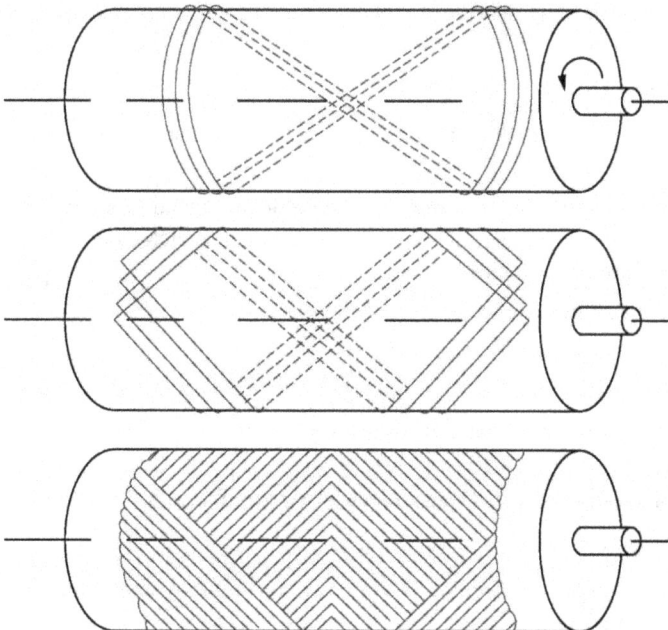

FIGURE 15.27 Filament winding pattern.

These structures are designed for specific load conditions, such as internal or external pressure in the case of tanks and pipes, and bending and compression in the case of masts for sailing vessels, etc.

15.9.2   FILAMENT WINDING STEPS

- Mandrel Preparation: A release agent or a thin plastic wrap is applied to the mandrel to facilitate easy removal of the finished part.
- Fibre Selection and Impregnation: The commonly used fibres include carbon, glass, or aramid. They are impregnated with a resin matrix (epoxy, polyester, or vinylester).

  For impregnation, a "wet winding" method can be used, where fibres are passed through a resin bath before winding. A prepreg winding can also be used, where pre-impregnated fibres are applied with the resin already on them.
- Controlled Pattern: Fibres are wound around the mandrel in a controlled manner, using specific angles for strength optimisation.
- Fibre Tensioning: Tension control ensures proper consolidation and alignment of fibres.
- Curing: The wound composite is cured to harden the resin and lock the fibres in place. The curing methods include room temperature for some thermoset resins or elevated temperature applications using ovens or autoclaves for high-performance results.
- Mandrel Removal: After curing, the mandrel is removed.
- Trimming and Finishing: Excess material is trimmed, and the part is inspected for defects. Additional coatings or treatments may be applied.

15.9.3   FILAMENT WINDING DESIGN VARIABLES

- Material selection: A wide range of fibres and matrices can be used for this process.
- The viscosity of the resin must not be too low to avoid problems in the final product. Viscosity should not be too high either, which can result in fraying of the fibre and bubbles.
- Winding angle around the axis of the mandrel.
- Reinforcement winding tension.
- Fibre volume content.
- Number of layers to reach the design thickness.

15.9.4   ADVANTAGES AND DISADVANTAGES OF THE FILAMENT WINDING PROCESS

- Advantages
- High Strength to Weight Ratio: Tailored fibre orientation maximises performance.

- Customizable Design: Fibre angles and resin types can be adjusted for specific mechanical properties.
- Cost Efficiency: Minimal material wastage and efficient production for large, hollow parts.
- Suitable for small to very large structures, including pipelines and rocket casings, with high-quality finishes and minimal voids.
- Disadvantages
- Geometry Limitations: Primarily suitable for symmetrical, cylindrical, or spherical shapes.
- High Initial Cost: Requires specialised equipment and mandrels.
- Labour-Intensive: Setup and mandrel removal can be time-consuming.

15.10   RIM (REACTION INJECTION MOULDING)

Reaction Injection Moulding (RIM), unlike traditional injection moulding, is a manufacturing process where chemical reactions are involved within the mould to form the final product.

To proceed with RIM, a closed mould is used, with the particularity that the matrix is in the form of a reactive two-component resin system, where the reinforcements are already premixed into one of these components.

As shown in Figure 15.28, prior to injection, the mould is closed and the two reactants are allowed to mix in the mixing head. Each component is constantly

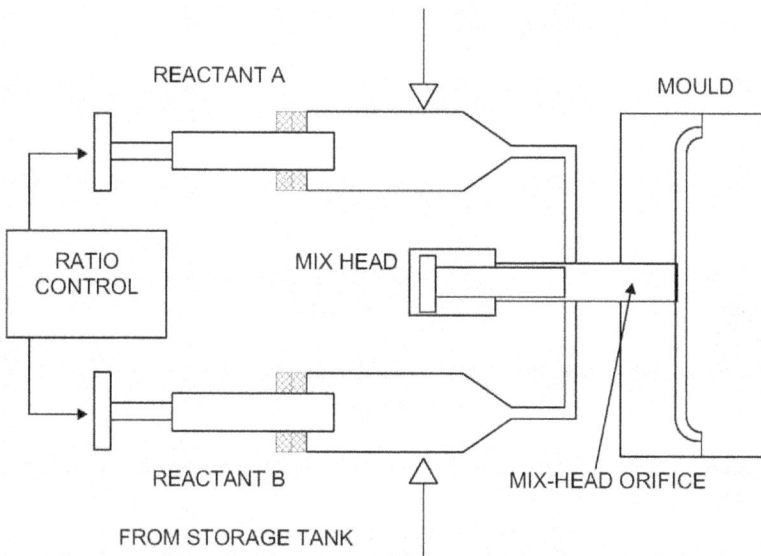

FIGURE 15.28   RIM injection moulding diagram.

recirculating on an independent path through a mixing head at low pressure, making the material flow into the mould until it cures.

The RIM process is comprised:

- Fibres: short glass (fibreglass).
- Matrices: thermosetting, usually polyurethane (which can be formulated to give flexible, rubbery materials or rigid, glassy materials).

RIM is slightly different from the normal injection moulding processes. However, it is a functional, practical process that suits the modern plastic injection moulding industry.

15.10.1 ADVANTAGES OF RIM

- Complex shapes can be produced with high precision.
- Lightweight components with high strength are achievable.
- Incorporates reinforcements such as glass or carbon fibres for improved properties improved.

15.10.2 APPLICATIONS

- Automotive body panels, bumpers, and interior parts.
- Medical device housings.
- Structural components in aerospace and furniture.
- Building panels.
- Casting for electrical equipment.
- Medical implants.

15.11 SPRAY-UP MOULDING

Spray-up moulding is an evolution of the hand lay-up lamination process, which has already been covered. It is an open-moulding technique popular for producing large, lightweight parts with relatively low tooling costs. It is particularly suited for fiberglass-reinforced polymer (FRP) composites. These days, it is automated for better productivity and lower manufacturing costs.

The lamination process is identical to the hand lay-up; the difference being that in this method, a spray gun is used to apply a mixture of catalysed liquid resin (polyester or epoxy) and chopped fibres (typically fibreglass) together onto the mould surface (viscosity between 300–500 cPs at 25°C). Figure 15.29 shows a representation of the spray-up moulding applied into a hull female mould. This is possible with the use of a special spraying equipment called a "spray-up gun" (Figure 15.30), so that the matrix and the reinforcement are applied simultaneously to the mould. The spray gun chops continuous fibre strands into short lengths (around 0.5–2 inches) and combines them with the resin before spraying.

The operator sprays multiple layers to achieve the desired thickness and structural properties. Additionally, layers of continuous fibre mats or fabrics can be hand laid-up between sprays for reinforcement if needed. This process must be

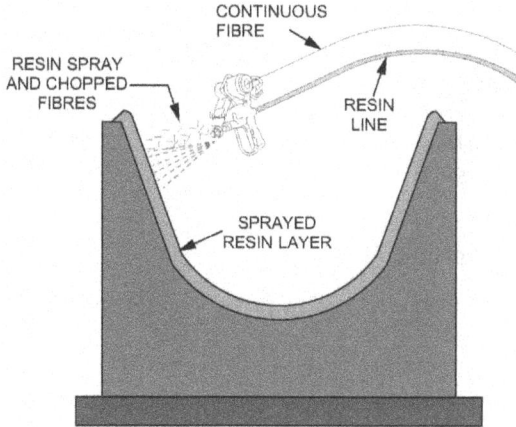

FIGURE 15.29 Representation of spray-up moulding on a hull mould.

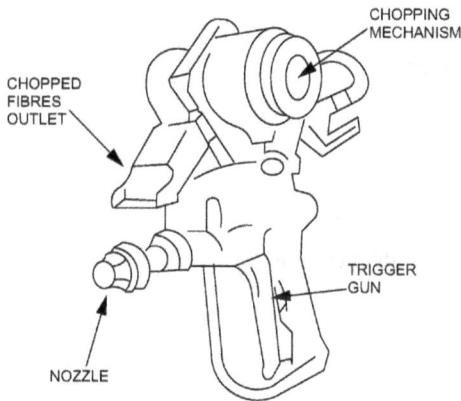

FIGURE 15.30 Spray-up gun.

performed with the correct proportions and in such a way that the quantities applied create a homogeneous layer in terms of thickness and distribution. Once applied, a compacting process must be done, using metallic consolidation rollers or squeegees to improve the reinforcement impregnation and eliminate air bubbles trapped between layers. This is critical for achieving a strong, void-free composite part.

15.11.1 SPRAY-UP MOULDING STEPS

- Mould tool preparation
- Reinforcement preparation
- Machine setting

- Spray-up
- Reinforcement impregnation and consolidation
- Material polymerization
- Demoulding
- Deburring and trimming
- Quality control (QC).

15.11.2 MACHINE SETTINGS

There are certain adjustments that allow proper control and stability of the spray-up manufacturing process:

- **Fibre flow:** The cutter setting is adjusted by approximately locating the pressure value of the cutter manometer and verifying the fibre flow by collecting and checking the weight of the cut filaments for one minute.
- **Resin flow:** This is adjusted depending on the amount of cut fibreglass, so that the weight ratio between the fibres and the resin is between 25 and 35 percent.
- **Flow rate of the catalyst system:** This adjustment is straightforward since, in most machines, the catalyst pump is shared with the resin pump, having an identical ratio for the pumped volumes.

15.11.3 ADDITIONAL SPRAY-UP SETTINGS

- Length of the cut filaments: Lengths between 25 and 50 mm.
- Blade condition.
- Rubber roller condition.
- Roller system adjustments (distances).
- Resin direction: Select the appropriate nozzle (check each manufacturer's specifications).
- Filaments direction: Select the appropriate nozzle (angle and nozzle diameter must be correct).

15.11.4 ADVANTAGES AND DISADVANTAGES OF THE SPRAY-UP PROCESS

- Advantages
- Low tooling costs compared to closed moulding methods
- Suited for large, complex shapes
- Flexible design options with various fibre orientations
- Relatively quick production process
- Disadvantages
- Limited control over fibre content and distribution
- Surface finish may not be as high-quality as other methods like compression moulding
- Specialised skilled labour-dependent process.

15.12 COMPRESSION MOULDING

The compression moulding process is a high-pressure, closed-mould manufacturing technique widely used to produce composite materials by shaping them into a desired form using heat and pressure within a mould. As shown in Figure 15.31, the material is placed into a mould cavity, and a shaped part is created by the application of heat and high pressure. It is commonly used for thermosetting polymers but can also be applied to thermoplastic composites. It is a low-cost process compared to the injection moulding and transfer moulding processes. This manufactured part will conform to the shape of the mould.

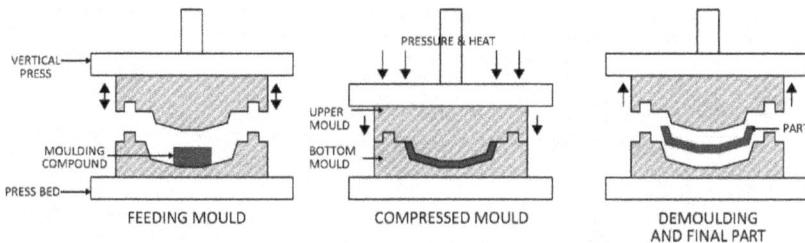

FIGURE 15.31 Compression moulding process.

In the compression moulding of thermosets, the mould remains hot throughout the entire cycle; as soon as a moulded part is ejected, new material can be introduced to start the process again.

On the other hand, unlike thermosets, thermoplastics must be fully cooled to harden. For thermosets, once the mould and the part have been cooled down below the gel point temperature, it is safe for demoulding and starting again. So, basically, before a moulded part is ejected, the entire mould must be cooled down, and as a result, the process of compression moulding could be quite slow.

15.12.1 COMPRESSION MOULDING PROCESS STEPS

15.12.1.1 Material Preparation

The composite material is prepared and cut to the required size.

15.12.1.2 Mould Setup

The mould, usually made of steel or aluminium, is preheated to the required temperature (typically 150–200°C for thermosets and higher for thermoplastics).

15.12.1.3 Placing Material into the Mould

The prepared composite material is placed inside the lower mould or tooling.

15.12.1.4 Compression and Heating

The upper mould is then closed, and high pressure is applied (from 5 to 100 MPa) using a hydraulic press.

Heat and pressure push the resin to flow and fully impregnate the reinforcing fibres while curing (for thermosets) or melting (for thermoplastics).

15.12.1.5 Cooling and Curing

For thermosets, the material cures into its final rigid state. For thermoplastics, the part is cooled under pressure to solidify. The cooling process can take anywhere from seconds to minutes, depending on the material and part size.

15.12.1.6 Demoulding and Trimming

Once the composite part has fully cured or solidified, it is removed from the mould.

Excess material is trimmed, and additional finishing operations, such as painting or drilling are performed.

15.12.2 ADVANTAGES AND DISADVANTAGES OF COMPRESSION MOULDING PROCESS

- Advantages
  - Fast initial setup
  - Good surface finish of the moulded parts.
  - Waste of material is considerably reduced compared with other processes.
  - Thermoplastic composites with unidirectional tapes, woven fabrics, randomly oriented fibre mat, or chopped fibres can be manufactured.
  - Consistency and Precision: Ideal for producing high-volume parts with repeatability.
- Disadvantages
  - Low production rate.
  - The moulds or the parts are limited to flat or slightly curved shapes, avoiding any undercuts or complicated shapes that make releasing the manufactured part difficult.
  - High initial tooling costs

15.13 PULTRUSION PROCESS

Pultrusion is a continuous manufacturing process used to produce composite materials with constant cross-sectional profiles. It is particularly well-suited for creating long, strong, and lightweight structural components, such as beams, rods, and tubes.

In a pultrusion process, fibres are pulled from the fibre roving rolls through a resin impregnation tank (additives, fillers, or pigments may be mixed with the resin for specific properties). The resin-impregnated fibres pass through a pre-forming guide, which aligns and compacts them into the desired shape, and then are pulled through a heated steel die where they are shaped and cured simultaneously. The heat initiates the polymerization of the resin, solidifying the composite.

Curing die: Completes the impregnation of fibres, controls the resin content, and cures the material to achieve the final shape.

Once this cured profile is fully polymerized, it passes to the final stage where it is cut/chopped to the desired length.

Fabrics can also be introduced into the die to provide directionality other than 0 degrees. Although pultrusion is a continuous process, producing a profile of constant cross-section, a simple variant of pultrusion known as "pulforming" allows for some variation to be introduced into the mentioned cross-section.

Material options include generally any epoxy, polyester, vinylester, or phenolic resin combined with any fibre type. Core materials are not generally used.

15.13.1 Advantages and Disadvantages of the Pultrusion Process

- Advantages
  - Can be a very fast production method, and as a result, very economical. This is mainly because of the impregnating and curing material process.
  - Resin content can be controlled accurately.
  - Fibre cost is reduced by the benefit of using the material straight from the rolls, resulting in less waste.
  - It can achieve good structural properties in the laminates due to the direction of the fibres through a long profile and the high volume of fibre content.
- Disadvantages
  - The process or the part to be manufactured is limited only to profiles/components with constant or near-constant cross-section.
  - Costs for introducing heat in the dies can be high.

15.13.2 Pultrusion Applications

Typical applications include beams and girders used in roof structures, bridges, ladders, and frameworks. Other applications such as

- Solid rods.
- Tubing.
- Long flat sheets.
- Structural sections (such as channels, angled and flanged beams).
- Tool handles for high-voltage work.
- Third rail covers for subways.

15.14 PULTRUSION PROCESS WITH FIBRE AND TAPE PLACEMENT

Pultrusion is a well-established method for producing continuous, constant cross-section profiles, but integrating fibre and tape placement adds a layer of customisation and performance improvement. By combining traditional pultrusion with fibre or tape placement, the mechanical properties of the composite can be optimized for specific applications. Figure 15.32 represents a fully automated pultrusion process diagram, where continuous fibres are pulled through a resin bath for impregnation, then shaped in a preform guide, pre-heated and then passing through a heated die to cure and form the final profile. After cooling, the cured composite is pulled by rollers and cut to length.

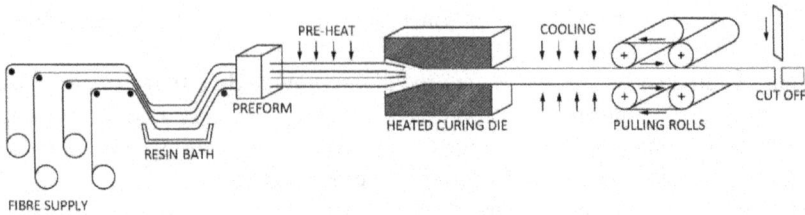

FIGURE 15.32 Pultrusion process diagram.

15.14.1 Fibre Placement in Pultrusion

- **Unidirectional Fibre Placement:** Aligning fibres along the length of the part maximises tensile and compressive strength in the longitudinal direction.
- **Multi-directional Fibre Placement:** Layers of fibres are arranged in multiple directions (±45°, 0°, 90°) for improved shear and transverse strength.

15.14.2 Tape Placement

Prepreg (pre-impregnated) tapes are used to strategically reinforce areas or layers of the composite. Tapes offer precise fibre alignment and resin content control, improving mechanical properties and minimizing voids.

15.14.3 Fibre and Tape Integration in Pultrusion

- Material Preparation.
- Layer Stacking: Tapes and fibres are laid in a specific sequence, either manually or via automated systems, before entering the die.
- Orientation can vary from unidirectional layers if strength is required, or cross-ply/angle-ply layers if balanced properties are needed.
- Shaping and Curing: The material stack is pulled through a heated die. Heat and pressure cure the resin matrix and consolidate the composite structure.
- Cutting and Post-Processing: The pultruded profile is cut to the desired length and undergoes finishing processes as needed.

15.14.4 Advantages of Fibre and Tape Placement in Pultrusion

- Advantages
  - Good mechanical properties
  - High precision
  - Lightweight and efficiency
  - Customisation.

TABLE 15.1

Summary of different manufacturing methods for composite materials

Manufacturing Method	Description	Advantages	Disadvantages	Outcome	Applications	Cost	Typical Materials Used
Hand lay-up	Layers of resin and fibre manually placed in a mould and cured	• Low setup cost • Simple process • Customizable thickness	• Labor-intensive • Risk of voids • Limited precision	Basic quality, suitable for large parts	Boats, tanks, wind turbine blades	Low	Glass fibres, polyester resin, epoxy
Vacuum Bagging Hand Lay-Up	Hand lay-up process combined with a vacuum bag to remove air and compact the laminate	• Improved fibre-to-resin ratio • Reduced voids • Better strength and finish	• Requires vacuum equipment • Time-consuming	Improved quality reducing defects	Aerospace, marine, sporting goods	Moderate	Glass or carbon fibres, epoxy resin
Pre-impregnated laminates	Pre-impregnated fibre reinforcements placed in a mould and cured, often under heat and pressure	• High consistency • Reduced resin waste • High strength	• Requires refrigerated storage • High initial cost	High-performance parts with excellent finish	Aerospace, automotive, high-performance goods	High	Prepreg carbon fibre, epoxy
Liquid transfer moulding (LTM)	Resin is injected into a closed mould containing dry fibres, with low-pressure methods	• Good surface finish • Low-cost tooling • Reduced voids	• Limited to small to medium parts • Limited fibre volume fraction	Good-quality for medium sized components	Wind energy, marine, automotive	Moderate	Glass or carbon fibres, polyester, epoxy resin
Resin transfer moulding (RTM)	Resin injected into a closed mould containing dry fibre reinforcement under higher pressure	• High surface finish • Accurate resin distribution • Moderate automation	• High tooling cost • Limited to medium sized parts	Strong and consistent parts	Automotive, aerospace, industrial components	High	Carbon fibre, glass fibre, epoxy resin
Infusion	Dry fibre reinforcement placed in an open mould, and resin is injected through under vacuum	• Low void content • Suitable for large parts • High fibre-to-resin ratio	• Slow process • Requires high leak inspection setup	Strong and lightweight components	Wind blades, marine, aerospace	Moderate to High	Glass or carbon fibres, epoxy resin

(Continued)

TABLE 15.1 (Continued)
Summary of different manufacturing methods for composite materials (Continued)

Method	Description	Advantages	Limitations	Best for	Applications	Cost	Materials
Autoclave moulding	Prepreg laminates cured under heat and pressure in an autoclave chamber	• Superior strength • Low void content • High precision	• Very high cost • Long cycle times	High-performance and quality components	Aerospace, military, high-end automotive	Very high	Prepreg carbon fibres, epoxy resin
Filament winding	Continuous fibres are wound around a mandrel, with resin previously applied, and cured	• High strength-to-weight ratio • Automated process • Excellent for rotational shapes	• Limited to symmetrical parts • Expensive equipment	Strong cylindrical or spherical components	Pressure vessels, pipes, rocket casings	High	Glass or carbon fibres, epoxy resin
RIM (Reaction Injection Moulding)	Two reactive liquid resins mixed and injected into a mould, reacting to form the composite.	• Fast cycle times • Good for large volume production • Automated process	• Limited fibre content • Lower mechanical properties than other methods	Economical parts with moderate quality	Automotive, appliances, electronics	Moderate	Polyurethane, glass-filled polymers
Spray-up moulding	Chopped fibres and resin are gun sprayed into an open mould	• Low-cost setup • Fast for large parts • Low material waste	• Poor fibre alignment • Limited strength	Low-strength, cost-effective parts	Bathtubs, panels, vehicle components	Low	Chopped glass fibres, polyester resin
Compression Moulding	Composite material placed in a two parts mould and compressed under heat and pressure to gain final shape.	• High precision • Fast cycle times • Consistent quality	• High tooling cost • Limited to small parts	High-quality, uniform parts	Automotive, consumer goods, electronics	High	Sheet moulding compounds (SMC), BMC
Pultrusion	Continuous fibres pulled through resin and a heated die to form constant cross-section parts.	• High production speed • Low waste • Excellent for long parts	• Limited to constant cross-sectional shapes • Expensive setup	Long, straight, lightweight components	Beams, rods, ladders, window frames	Moderate	Glass or carbon fibres, polyester, vinylester

15.15   SUMMARY TABLE OF DIFFERENT MANUFACTURING METHODS

To simplify what has been shown through all the composite manufacturing methods explained throughout this chapter, Table 15.1 summarised a detailed comparison outlining common manufacturing methods used for composite materials.

BIBLIOGRAPHY

'Analysis and Performance of Fiber Composites' (Third edition). Bhagwan D. Agarwal, Lawrence J. Broutman and K. Chandrashekhara. Wiley India Pvt. Ltd., 2015.

'Composite Manufacturing Process Selection Using Analytical Hierarchy Process'. A. Hambali, S. M. Sapuan, Napsiah Ismail and Yusoff Nukman. *International Journal of Mechanical and Materials Engineering (IJMME)*, Vol. 4 (2009), No. 1, 49–61.

'Composite Manufacturing Technology (Soviet Advanced Composites Technology Series, 1)' (1994th edition). A. G. Bratukhin and V. S. Bogolyubov. Dordrecht: Springer Science+ Business Media, 1994.

'Composite Materials'. S. C. Sharma. Alpha Science International, Ltd, 2000.

'Composite Materials: Design and Applications'. Daniel Gay, Suong V. Hoa and Stephen W. Tsai. CRC Press LLC, 2003.

'Engineering Mechanics of Composite Materials' (Second edition). Isaac M. Daniel and Ori Ishai. New York: Oxford University Press, 2006.

'Essentials of Advanced Composite Fabrication & Repair' (Second edition). Louis C. Dorworth, Ginger L. Gardiner and G. M. Mellema. Aviation Supplies & Academics, 2019.

'Handbook: An Engineering Compendium on the Manufacture and Repair of Fiber Reinforced Composites'. R. L. Ramkumar, N. M. Bhatia, J. D. Labor and J. S. Wilkes. NJ, USA: Department of Transportation FAA Technical Center: Atlantic City International Airport, 1987.

'Handbook of Composites' (Second edition). S. T. Peters. Mountain View, CA, USA: Process Research, 1997.

'Introduction to Composite Materials Design'. Ever J. Barbero. USA: Department of Mechanical & Aerospace Engineering – West Virginia University/Taylor & Francis, 1998.

'Marine Composites' (Second edition). Eric Green Associates. MD: Eric Green Associates Inc., 1999.

'Mechanics of Composite Materials' (Second edition). Robert M. Jones. Taylor & Francis, 1999.

'Resin Transfer Moulding for Aerospace Structures' (1998th edition). T. Kruckenberg and R. Paton. Springer, 2012.

'Structural Composite Materials'. F. C. Campbell. ASM International, 2010.

16 Factors to be Considered in the Manufacturing Process Selection

The right or most convenient selection of the manufacturing process depends on a huge number of factors, and most of these factors are interrelated.

All processes have their own strengths, weaknesses, and priority in selection. Getting the right manufacturing process is a complex activity, where following a correct methodology will help make the approach to the process selection more systematic. However, the most optimum manufacturing process must be selected according to the design requirements and factors that suit best for our design.

To determine the most appropriate manufacturing process during the early design stage, it is important to pay attention to the details of each of these factors, which include

- Dimensions and geometry
- Production characteristics and volume
- Material consideration
- Cost consideration
- Maintenance
- Equipment and labour availability.

16.1 DIMENSIONS AND GEOMETRY

The dimensions and geometry of a composite part have an important influence on process selection, based on the need for precision, structural performance, surface quality, and cost efficiency. Ensuring alignment of the selected process with these design aspects is critical for producing high-quality composite components.

The selection of a suitable manufacturing process could be influenced by the geometry of the design, and many factors related to the geometry of the design need to be considered:

a) **Shape:** The shape of the product is the most important factor that must be considered when determining the most suitable manufacturing process. As more complex the shape is, the selection of a suitable process becomes quite important.

b) **Design complexity:** Complexity is defined as the presence of design features such as non-uniform wall thickness, non uniform cross section, and holes. These design features have to be considered to avoid any additional

 DOI: 10.1201/9781003565222-16

processes and increase production time during the manufacturing process. It is also the designer's job to avoid any possible modification of the geometry to match up with the chosen manufacturing process.

c) **Size:** The size of the designed part required to be manufactured is an important factor that needs to be considered. Depending on each process, there will be limitations.

d) **Wall Thickness:** The wall thickness is also an important factor to consider, which will influence the selection of a suitable manufacturing process. For thin parts, processes with uniform resin distribution are required, such as vacuum bagging or compression moulding. However, thick parts will need processes like resin infusion or RTM to ensure proper curing and minimise voids in the material.

e) **Surface finish:** The surface finish will be managed by the measured roughness or smoothness of the surface on the final product. Processes like prepreg lay-up, compression moulding, or RTM provide excellent surface finishes with minimal post-processing, while hand lay-up or spray-up might be used when aesthetics are not a primary concern.

16.2   PRODUCTION CHARACTERISTICS

Production characteristics are a very essential factor in getting the most suitable manufacturing process. There are three production characteristics that may influence the selection of the correct manufacturing process:

1) **Production volume:** Production volume plays an important role in the manufacturing process selection. This will determine whether or not a manufacturing process is suitable for the required purposes.

2) **Production rate:** The right selection of the manufacturing process is also based on the rate of production. Each process has its own possible production rate.

3) **Processing times:** For a high volume of production, it is important to reduce the processing times; this will also affect the cost considerably.

16.3   MATERIAL CONSIDERATIONS

The selection of the manufacturing process is directly related to material selection. It is convenient that the material has been selected at the early stage of the product development process since the design will also be related to this factor, especially considering material properties, thicknesses, dimensions, and geometry. So, at some point during the design process, it is important to think about all of these factors at the same time to develop a proper product.

These considerations ensure compatibility between the composite's matrix and reinforcement materials, as well as their alignment with process capabilities and their requirements for usage.

16.4   COST CONSIDERATIONS

It is well known that costs are an important factor to consider; in different words, any production parameter is related to cost. They are critical to ensuring economic viability while meeting performance and production requirements.

To achieve a proper final cost, several factors must be considered and can be divided into material costs, tooling costs, equipment costs, labour costs, and post-processing costs:

a) **Material cost:** Choosing the right combination of matrix and reinforcement materials, optimizing waste, and selecting the appropriate scale of production can help achieve cost-effective and high-performance composite components.

b) **Tooling cost:** The tooling cost is very important. It will depend on the type of processes to be used, design complexity, and the volume of production. There are ways to reduce the tooling cost, but it will affect the final quality of the product, such as surface finish or labour times.

c) **Equipment cost:** Lower equipment or machine cost is also an important factor that needs to be considered in the selection of the manufacturing process; this will also include energy consumption and cycle time.

d) **Labour cost:** This factor is also managed by the tooling and process selection since a complex manufacturing process will require more specialised labour experience. At the same time, if the manufacturing process requires a certain number of process stages, this will increase the cost as well.

e) **Post-processing cost:** Post-processing costs are influenced by the chosen manufacturing process, material properties, and part complexity. Selecting the right process and optimizing tooling, automation, and material use can significantly reduce post-processing costs.

16.5   MAINTENANCE

Maintenance is a crucial factor in the selection of a manufacturing process for composite materials, as it directly impacts operational efficiency, equipment longevity, and overall production costs. The right choice of a manufacturing process is also considered based on how often the need for tooling, tools, or systems maintenance is required. Sometimes, products fail due to the lack of maintenance.

16.6   EQUIPMENT AND LABOUR AVAILABILITY

The availability of equipment and labour is a key factor in determining a manufacturing process. Processes with low equipment and labour requirements, like hand lay-up, are suitable for productions with limited resources. On the other hand, automated processes that include advanced machinery and skilled technicians, such as injection moulding, are ideal for high-volume production.

The availability of equipment and labour is also an important factor; not having the proper availability will result in delays in the product.

16.7 METHODOLOGY OF MANUFACTURING PROCESS SELECTION

The production process selection flowchart shown in Figure 16.1 will facilitate the view of factors and sub-factors that are involved in the process selection.

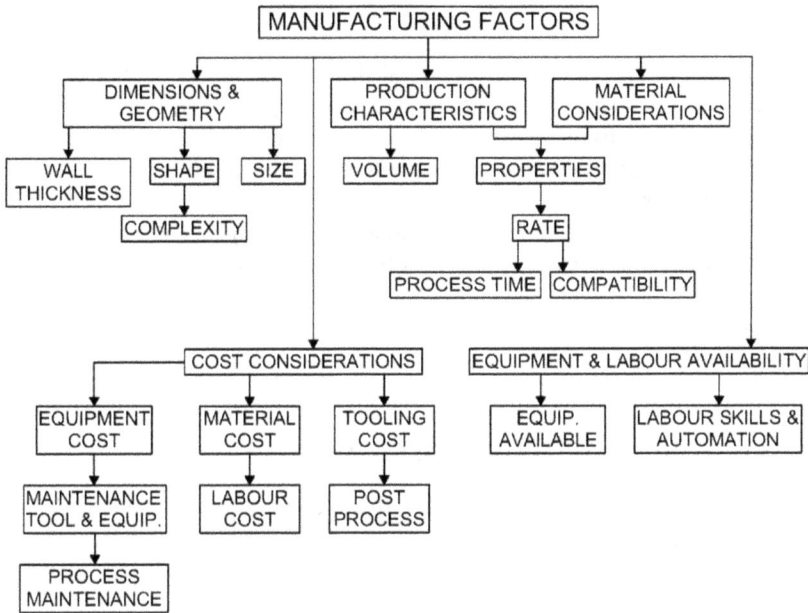

FIGURE 16.1   Production process selection chart.

BIBLIOGRAPHY

'Analysis and Performance of Fiber Composites' (Third edition). Bhagwan D. Agarwal, Lawrence J. Broutman and K. Chandrashekhara. Wiley India Pvt. Ltd., 2015.

'Composite Manufacturing Process Selection Using Analytical Hierarchy Process'. A. Hambali, S. M. Sapuan, Napsiah Ismail and Yusoff Nukman. *International Journal of Mechanical and Materials Engineering (IJMME)*, Vol. 4 (2009), No. 1, 49–61.

'Composite Materials: Design and Applications'. Daniel Gay, Suong V. Hoa and Stephen W. Tsai. CRC Press LLC, 2003.

'Introduction to Composite Materials Design'. Ever J. Barbero. USA: Department of Mechanical & Aerospace Engineering – West Virginia University/Taylor & Francis, 1998.

17 Composite Product Design

Designing products with composite materials involves a careful balance of material properties, manufacturing capabilities, and product requirements.

Designers of composite parts can choose from a huge variety of fibre reinforcements and resin systems; this creates infinite design freedom but also adds some complexity to composites.

To help understand the process of designing a product made with composites, two sets of requirements can be set as the basis for these decisions:

- Material selection.
- Manufacturing process selection.

17.1 STEPS IN COMPOSITE PRODUCT DESIGN

- Define Requirements: Mechanical, thermal, and environmental performance criteria.
- Preliminary Design: Concept sketches where critical dimensions and features are set.
- Material and Process Selection: Related to cost-performance.
- Prototype Development: Building and testing prototypes using chosen materials and methods will help in the process.
- Testing and Validation: Verifying performance through mechanical, thermal, and environmental testing.
- Production Design: Incorporate the latest updates based on previous steps and refine the design for manufacturing and production.

17.2 MATERIAL SELECTION

Setting and knowing the material properties is a key point in product design, but cost is a major factor as well. In this way, overdesigned composites cannot compete with lower-cost composites.

In any composite design, the fibre reinforcements will provide the main portion of the composite's mechanical properties, such as stiffness and strength, while the matrix will provide physical characteristics such as toughness and resistance to impact, weather, ultraviolet (UV) light, and/or corrosive materials.

DOI: 10.1201/9781003565222-17

17.2.1 MATERIAL DESIGN CONSIDERATIONS

First, it is important to remind that the fibres provide the primary strength of the composite. There are additional factors that must be considered when designing with fibre:

- **Fibre Orientation and Lay-up:** The use of unidirectional, woven, or randomly oriented fibres will be directly related to the load requirements. Optimizing the lay-up sequence is an important consideration to avoid warping and delamination.
- **Matrix Selection:** Choosing a matrix material should be guided by selecting the best complement that suits the fibres and also meets environmental resistance requirements. Ensuring compatibility between the matrix and reinforcement is a key consideration to gain efficient load transfer.
- **Thickness and Layer:** It is important to find a balance between stiffness and weight, considering manufacturing limitations. Avoid excessive layers, which could lead to voids or incomplete curing.
- **Cure Cycle Optimisation:** Heat and pressure requirements need to be considered for thermoset vs. thermoplastic composites. The use of autoclaves or ovens with controlled environments will provide a consistent quality.

17.3 MANUFACTURING PROCESS SELECTION

The fabrication method will also influence the design; integrating manufacturing process selection into composite product design is a critical step that directly influences the product's performance, cost, and feasibility. For example, manufacturers of filament-wound structures use different reinforcement forms and patterns than those in hand lay-up or autoclave prepreg laminates. Resin transfer moulding (RTM) uses a fully enclosed mould and requires three-dimensional preforms in comparison with other techniques. In addition, the choice of production procedures and type of tooling must be decided together with the part geometry, part service requirements, and the quantity of production demand.

17.3.1 FACTORS TO CONSIDER IN THE MANUFACTURING PROCESS SELECTION

The choice of a manufacturing process will depend on:

- Product Complexity: Geometries, features, and dimensions.
- Volume Requirements: Prototypes, low volume, or mass production.
- Material Requirements: Reinforcement type (fibres, particles) and matrix type (thermoplastics, thermosets, or hybrids).
- Mechanical Performance: Structural loads, impact resistance, and anisotropic properties.

- Cost Constraints: Equipment, tooling, labour, and material costs.
- Surface Finish Requirements: Aesthetics or functional needs like smoothness or uniformity.
- Part Size: Larger parts may require specialised methods, such as filament winding or resin infusion.

17.3.2 STEPS TO INTEGRATE PROCESS SELECTION

- Define Design Requirements: Set performance criteria, dimensional tolerances, and material preferences.
- Matching Process Capabilities: Compare the requirements with each manufacturing process's capabilities. For example, for high-volume automotive parts, compression moulding may be optimal, while for large structural aerospace parts, RTM might be better.
- Evaluate Cost and Efficiency: Estimate/calculate the tooling costs, cycle times, and labour requirements for each process.
- Prepare Prototype using a Suitable Process: Start with a lower-cost process, like hand lay-up or 3D printing, to evaluate design feasibility. Then, the transition to production processes could be done once designs are validated.
- Design for the Process: Adjust part geometry to fit the selected process. This refers to the use of details such as draft angles for demoulding in compression or RTM.
- Validate and Create Iterations: Perform mechanical and non-destructive testing on samples. Refine the design and process based on testing feedback.

17.4 STRENGTH AND STIFFNESS OF MATERIALS

The strength and stiffness of a composite material are related to the product design, and they will depend on the material's constituents (fibre and matrix), their arrangement, and the manufacturing process. These properties must be carefully tailored to meet the product's performance requirements.

17.4.1 MATERIAL STRENGTH IN COMPOSITE DESIGN

Strength refers to a material's ability to withstand applied forces without breaking. In composites, this relies on:

a) Fibre Properties:
  - Type of Fibres: Carbon fibres provide high tensile and compressive strength. Glass fibres are strong and cost-effective but less stiff than carbon. Aramid fibres offer high impact resistance but lower compressive strength.
  - Volume Fraction: Higher fibre content typically increases strength.

- Orientation: Unidirectional fibres maximise strength along the fibre direction. Woven fibres provide more isotropic strength but lower peak performance.

b) Matrix Properties:
- The matrix, such as epoxy or polyester, transfers stress between fibres. It contributes to compressive, shear, and interlaminar strength. Thermoplastics tend to improve toughness but may reduce overall strength compared to thermosets.

c) Fibre-Matrix Interface:
- Strong adhesion ensures efficient stress transfer.
- Poor bonding can lead to delamination or fibre pull-out.

d) Lay-up Configuration:
- Multidirectional lay-ups can handle complex loads.

17.4.2 MATERIAL STIFFNESS IN COMPOSITE DESIGN

Stiffness measures the resistance of the material to deformation under load. In composites, stiffness is led by:

a) Fibre Stiffness: Carbon fibres have the highest stiffness among common reinforcements. Glass fibres are less stiff, making them suitable for less rigid applications.

b) Directionality: Along the fibre direction, stiffness is dominated by the fibres. Perpendicular to fibres, stiffness is determined by the matrix and fibre volume fraction.

c) Fibre Volume Fraction (FVF): Higher FVF increases stiffness, especially in the fibre direction.

d) Laminate Thickness: Increased thickness in lay-ups improves flexural stiffness but may also increase weight.

e) Lay-up Sequence: Symmetric and balanced lay-ups reduce warping and ensure even stiffness distribution. Cross-ply laminates offer good in-plane stiffness.

17.4.3 RELATIONSHIP BETWEEN STRENGTH AND STIFFNESS

When people talk about the strength of materials, there is only one definition to describe it and this is the Young's modulus of elasticity (E), which is a measure of the stiffness of the material. It is defined as the slope of the linear portion of the normal stress-strain curve of a tensile test conducted on a sample of the material.

The behaviour of the composite under mechanical loading can be reflected in a stress-strain curve, as shown in Figure 17.1, influenced by the fibre, matrix, and fibre-matrix interaction.

In a stress-strain curve, yield strength, σy, and ultimate strength, σu, represent critical points that define material behaviour under mechanical loading. For composite materials, these properties are more complex than for isotropic materials due to their anisotropy and heterogeneous structure.

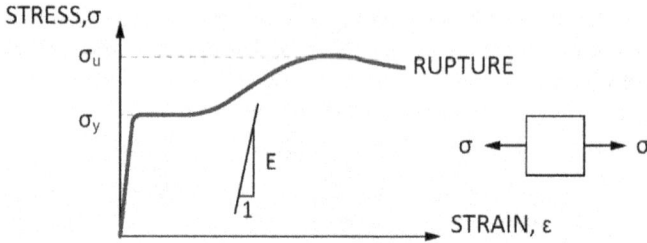

FIGURE 17.1   Stress-strain curve.

Definitions:

- **Yield Strength (σy):** Yield strength is the stress at which a material begins to deform plastically. Beyond this point, permanent deformation occurs.
- **Ultimate Strength (σu):** Ultimate strength is the maximum stress the material can withstand before failure. After this point, the material loses its ability to carry load.

The curve starts linear, deviates slightly if the matrix yields or cracks, and ends with a sharp drop or progressive failure. Looking at the plotted chart, the stress-strain curve for composites can be divided into different regions:

1) Initial Linear Elastic Region: Fibres carry most of the load in the loading direction, leading to a linear relationship between stress and strain. The matrix contributes to transverse and shear properties, but its stiffness is generally lower than the fibres. This region extends until the matrix starts to yield or micro-cracking occurs.
2) Non-linear Region: The matrix may start to yield; if the matrix begins to yield (in thermoplastics), the curve may exhibit slight non-linearity. Micro-cracking can occur, and they may form at fibre-matrix interfaces, particularly in transverse loading.
3) Failure Region: The first indication of failure will start progressively, which normally will be reflected in the matrix first. Delamination or fibre pull-out may follow.

Final fibre failure, in unidirectional composites for example, fibres fail abruptly when their tensile strength is exceeded. In woven or multidirectional composites, the failure can be gradual due to load redistribution.

17.4.4   NORMAL STRESS AND NORMAL STRAIN

Normal stress and normal strain are fundamental concepts in the mechanics of materials, describing how a material responds to axial forces.

Normal Stress (σ): Is the internal force per unit area acting perpendicular to a surface within a material. It leads to expansion or contraction.

Normal Strain (ε): Is the measure of deformation or elongation of a material per unit of its original length due to applied stress, the change in length over the original length, L.

BIBLIOGRAPHY

'Analysis and Performance of Fiber Composites' (Third edition). Bhagwan D. Agarwal, Lawrence J. Broutman and K. Chandrashekhara. Wiley India Pvt. Ltd., 2015.

'Analysis of Composite Materials. A Survey'. Z. Hashin. *Journal of Applied Mechanics*, Vol. 50 (1983), No. 3, 481–505.

'Composite Manufacturing Process Selection Using Analytical Hierarchy Process'. A. Hambali, S. M. Sapuan, Napsiah Ismail and Yusoff Nukman. *International Journal of Mechanical and Materials Engineering (IJMME)*, Vol. 4 (2009), No. 1, 49–61.

'Composite Materials: Design and Applications'. Daniel Gay, Suong V. Hoa and Stephen W. Tsai. CRC Press LLC, 2003.

'Handbook of Composites' (Second edition). S. T. Peters. Mountain View, CA, USA: Process Research, 1997.

'Introduction to Composite Materials Design'. Ever J. Barbero. USA: Department of Mechanical & Aerospace Engineering – West Virginia University/Taylor & Francis, 1998.

'Marine Composites' (Second edition). Eric Green Associates. MD: Eric Green Associates Inc., 1999.

'Mechanics of Composite Materials' (Second edition). Robert M. Jones. Taylor & Francis, 1999.

18 Fibre-to-Resin Ratio

The fibre-to-resin ratio defines the proportion of reinforcing fibres within the resin matrix in a composite material. It is a critical parameter that directly impacts the material's mechanical properties, weight, and overall performance. These mechanical properties depend on the combined characteristics of the fibre and resin, as well as the direction of the applied load.

Figure 18.1 is an example of how a volume of fibre and resin would look. As a quick overview, it can be observed that there is a lot of resin and not much fibre. Thinking of the fibres only, as a certain portion of the volume and putting it as a fraction of that volume, this will be considered and defined as "fibre volume fraction", which in this case is not much.

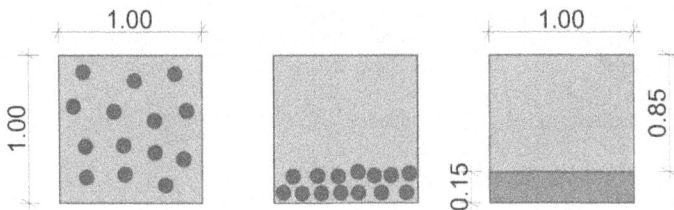

FIGURE 18.1   Representation of fibre-to-resin ratio on a composite material.

So, each property has influence on each volume in the composite's plies. And each ply has influence on the composite properties. To understand the fibre and resin volume ratios, it is necessary to know how much room each of them takes up in the ply. Knowing the density from the resin datasheets and the fibre weight, the total laminate weight can be estimated or predicted. But to estimate the volumes, first, the definition of what weight of the fibres means has to be explained.

18.1   WEIGHT OF THE FIBRES

The weight of the fibres in terms of the fibre volume fraction can be defined as the contribution of the fibre mass to the composite material, expressed as a function of the volume fraction of the fibres:

$$Wf = Vf \; x \; \rho f$$

where
- Wf: weight of the fibres (mass per unit volume of composite).
- Vf: fibre volume fraction (the volume of fibres divided by the total volume of the composite).
- ρf: density of the fibre material

                                        DOI: 10.1201/9781003565222-18

On the other hand, the weight of fibres could also be defined as the weight of a reinforcement material for a given area.

Example: Having a carbon woven reinforcement of 200 g/m² means that one square meter of that specific material will be equal to 200 g in weight.

Weight of Fibre = Weight of the reinforcement x Ply Area (m²)

The goal here is to be able to identify the values of fibre volume fraction and the fibre and resin weight ratios separately. So, the total weight could be calculated as shown in Figure 18.2

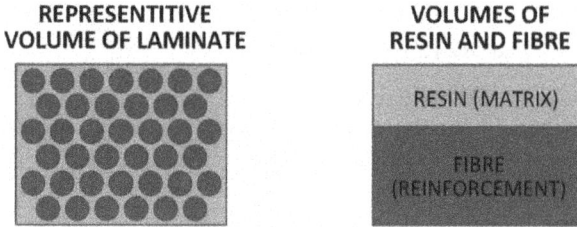

REPRESENTITIVE VOLUME OF LAMINATE

VOLUMES OF RESIN AND FIBRE

RESIN (MATRIX)

FIBRE (REINFORCEMENT)

EQUATIONS:

TOTAL WEIGHT= (RVF x RESIN DENSITY) + (FVF x FIBRE DENSITY)

FVF = 1 - (RVF x VOID FRACTION)

TOTAL WEIGHT = RESIN WEIGHT + FIBRE WEIGHT

FIGURE 18.2 Representation of the total weigh in a composite material.

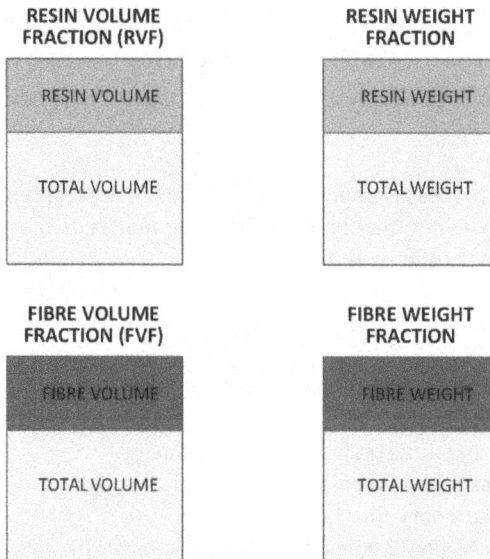

RESIN VOLUME FRACTION (RVF)

RESIN VOLUME

TOTAL VOLUME

RESIN WEIGHT FRACTION

RESIN WEIGHT

TOTAL WEIGHT

FIBRE VOLUME FRACTION (FVF)

FIBRE VOLUME

TOTAL VOLUME

FIBRE WEIGHT FRACTION

FIBRE WEIGHT

TOTAL WEIGHT

FIGURE 18.3 Representation of FVF, RVF, and resin and fibre weight fractions.

Taking as an example a stack of unidirectional fibres, the maximum fibre volume fraction (FVF) that could be obtained would be around 0.80.

18.1.1 TYPICAL FIBRE VOLUME FRACTIONS

- Open moulded hand lay-up: 20–40%
- Vacuum bagging wet lay-up: 40–55%
- Infusion: 40–55%
- Prepreg: 50–70%

18.1.2 TYPICAL FIBRE-TO-RESIN RATIOS

- Hand lay-up: ~30:70 (fibre:resin) by weight.
- Vacuum bagging: ~40:60 to 50:50
- Resin transfer moulding (RTM): ~50:50
- Autoclave processing: ~60:40
- Prepreg (pre-impregnated fibre): ~65:35 to 70:30

18.2 IMPORTANCE OF THE FIBRE-TO-RESIN RATIO

- Strength and Stiffness: A higher fibre content typically increases the strength and stiffness of the composite since the fibres carry most of the load. An insufficient fibre content can lead to reduced mechanical performance.
- Weight: Fibre is generally lighter than resin for many types of composites, so a higher fibre content often reduces overall weight.
- Durability and Toughness: Resin plays a significant role in protecting the fibres from environmental damage and distributing stress. Too little resin can leave fibres unprotected, while too much resin can make the composite brittle.
- Cost: Fibres are often more expensive than resin, so the ratio impacts material cost.

The manufacturing process is important as well; for example, prepreg in an autoclave will have a higher volume fraction than the same materials manufactured by hand lay-up compacted by a roller with low pressure.

18.3 LAMINATE THICKNESS ESTIMATION

Knowing the approximate fibre volume fraction, the laminate thickness can be estimated quite well. The higher the fibre volume fraction, the less resin content in the laminate would be, and as a result, a thinner laminate.

These estimations could be easily represented in a chart as a linear graph, with some exceptions, like a very thin laminate.

Laminate thickness could be estimated, for example, by looking at the graph in Figure 18.1, for a carbon at 0.5 FVF and 1000 g of fibre weight, the laminate thickness would be 11 mm.

LAMINATE THICKNESS BY WEIGHT, MATERIAL AND FIBRE VOLUME FRACTION

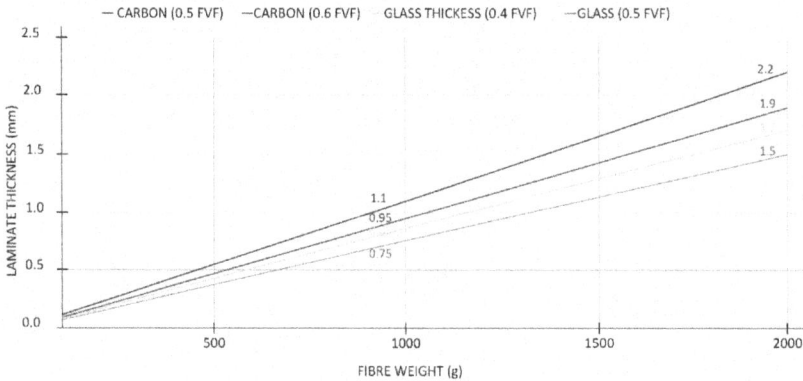

FIGURE 18.4   Representation chart of laminate thickness by weight.

The graph presented in Figure 18.4 is showing the relationship between laminate thickness, fibre weight, material type, and fibre volume fraction (FVF). Where the x-axis represents fibre weight (g), while the y-axis represents laminate thickness (mm). The chart includes multiple lines representing different materials and FVFs: carbon (0.5 FVF), carbon (0.6 FVF), glass (0.4 FVF), and glass (0.5 FVF). Each line shows how laminate thickness increases while increasing fibre weight.

$$t = \frac{n \, x \, Wf}{Vf \, x \, \rho f}$$

where
- t: total laminate thickness
- n: number of plies
- Wf: areal weight of the fibre (mass of fibres per unit area, typically in g/m²)
- Vf: fibre volume fraction (typically between 0.4 and 0.7 for structural composites)
- ρf: density of the fibre material in (g/cm³ or kg/m³).

Example Calculation

- Input Data:
  - Fibre areal weight (Wf): 300 g/m²
  - Fibre volume fraction (Vf): 0.6
  - Fibre density (ρf): 1.8 g/cm³
  - Number of plies (n): 8

$$t = \frac{8 \, x \, 300}{0.6 \, x \, 1800} = \frac{2400}{1080} \approx 2.22 mm$$

18.4   FIBRE DIRECTIONALITY

Fibre directionality in composite materials refers to the orientation of reinforcing fibres within the resin matrix. Properties of the composite will be strictly related to the direction of the fibres, where the mechanical properties of the fibres and their direction will have a huge influence on the final composite properties.

Looking at the curve presented in Figure 18.5 from the stiffness perspective, as the fibre angle changes, the fraction of maximum stiffness will drop.

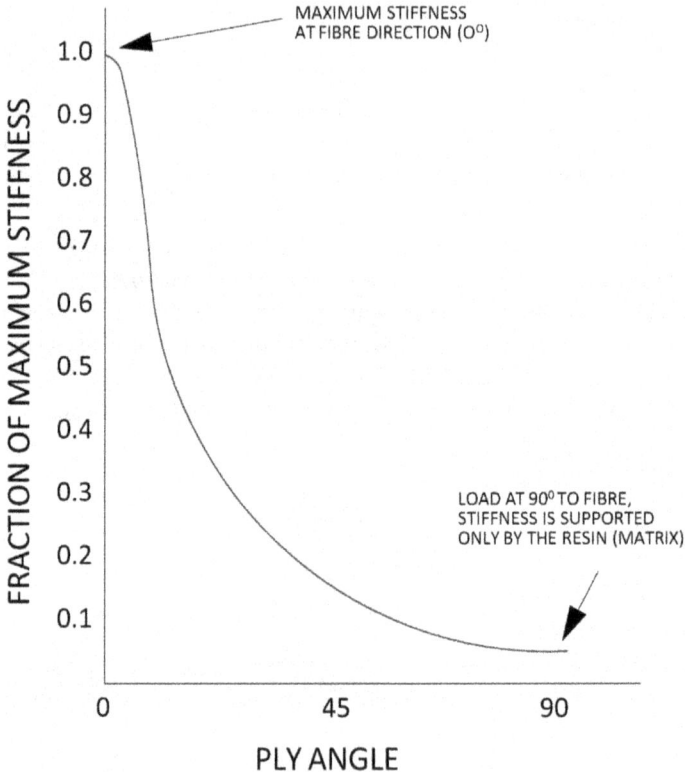

FIGURE 18.5   Representation chart of laminate thickness by weight.

As the load direction moves away from being aligned with the fibre direction, the stiffness drops rapidly, and when it reaches 90 degrees to the fibre direction, the stiffness will be purely based on the matrix stiffness.

18.4.1   EFFECTS OF FIBRE ORIENTATION ON PROPERTIES

- Strength: Maximum strength occurs along the fibre direction. Minimal strength in directions perpendicular to the fibres unless fibres are present in those directions.

- Stiffness: High performance directionality on stiffness, when aligned with fibre orientation. If loads are off-axis, then they will rely on matrix properties, which are weaker and less stiff.
- Failure Modes: Under longitudinal loading, there is a high load-carrying capacity. On the other hand, lower capacity, dominated by matrix properties, can be found under transverse loading.

Figure 18.6 is a typical "fibre orientation angle representation". These angles are normally relative to a reference axis, usually the x-axis of the global coordinate system or the principal loading direction.

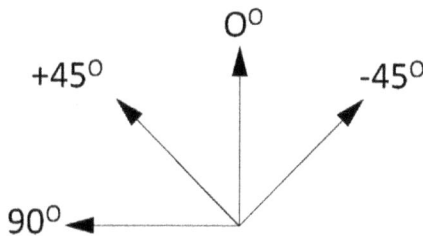

FIGURE 18.6   Typical representation of fibre orientation angles on different axes.

Fibre orientation angle coordinate systems are normally presented in laminate drawings to set up the parameters.

BIBLIOGRAPHY

'Analysis and Performance of Fiber Composites' (Third edition). Bhagwan D. Agarwal, Lawrence J. Broutman and K. Chandrashekhara. Wiley India Pvt. Ltd., 2015.

'Analysis of Composite Materials. A Survey'. Z. Hashin. *Journal of Applied Mechanics*, Vol. 50 (1983), No. 3, 481–505.

'Composite Manufacturing Process Selection Using Analytical Hierarchy Process'. A. Hambali, S. M. Sapuan, Napsiah Ismail and Yusoff Nukman. *International Journal of Mechanical and Materials Engineering (IJMME)*, Vol. 4 (2009), No. 1, 49–61.

'Composite Materials: Design and Applications'. Daniel Gay, Suong V. Hoa and Stephen W. Tsai. CRC Press LLC, 2003.

'Handbook of Composites' (Second edition). S. T. Peters. Mountain View, CA, USA: Process Research, 1997.

'Introduction to Composite Materials Design'. Ever J. Barbero. USA: Department of Mechanical & Aerospace Engineering – West Virginia University/Taylor & Francis, 1998.

19 Critical Variables during the Process Development in Composite Materials

Developing composite materials requires control and setup of critical variables across the different stages. These variables will play a key role in shaping the material's performance, manufacturability, and cost.

To summarize the different steps, here are the key aspects to consider during the product development process:

i) **Design and Performance Requirements**
- Product design
- Tooling design.

In this step, different properties will be analysed and defined, such as the mechanical properties (strength, stiffness, toughness, and fatigue resistance), thermal properties (thermal conductivity, heat resistance, and thermal expansion), chemical resistance (resistance to environmental degradation or exposure to chemicals), weight constraints, and durability (service life under specific environmental and loading conditions).

ii) **Material Selection**
This involves not just the product material where the product will be manufactured; matrix material needs to be set as well, including reinforcement type, fibre orientation, and interphase or coupling agents to ensure proper bonding between parts.

iii) **Process Selection**
Selecting the correct or most convenient process is very important, as well as the curing parameters, such as temperature and pressure.

iv) **Resin Selection**

In this step, it is important to keep in mind the resin characteristics and properties, as well as its different variables, such as resin flow and viscosity, which determine impregnation efficiency in fibres; fibre volume fraction, which directly influences mechanical properties; the lay-up sequence; and the cure cycle.

DOI: 10.1201/9781003565222-19

But there are also other steps to consider during the manufacturing process:

- **Material Control**
  - Material temperature < T resin degradation (temperature at which the resin starts to deteriorate).
  - Curing temperature.
  - Tg, which is the glass transition temperature.
- **Process Control**
  - Mould tooling temperature
  - Time
  - Pressure
  - Work environment.
- **Process Optimisation**
  - This can be achieved by enhancing efficiency, reducing waste, and improving product performance.

19.1 MATERIAL CONTROL

Material control is crucial to ensure consistent quality, performance, and reliability. Effective control methods focus on the materials themselves and the manufacturing processes.

There are three typical ways to test or control composite materials:

- Composite test specimens
- Destructive tests
- Non-destructive tests (NDTs)

19.1.1 Composite Test Specimen

A composite test specimen is a sample of a composite material that has been fabricated and prepared for testing to evaluate its properties under controlled conditions. The composite test specimen has to be taken from the manufactured part itself.

These specimens will be used for quality control tests; specifically, they are used to characterize the mechanical, thermal, chemical, and physical properties of composites, ensuring they meet the requirements for specific applications. As an observation, they can be fabricated in case of not having the chance to obtain them from the part itself.

19.1.1.1 Purpose of Composite Test Specimens
- To validate material properties for design and manufacturing.
- To assess the impact of processing techniques on performance.
  To ensure compliance with industry standards.

19.1.1.2   Factors to Consider for the Specimen Test

- Environmental conditions (specimen has to be exposed to the same environmental conditions).
- Material density (fibre volume and resin control).
- Environmental conditions (specimen has to be exposed to the same environmental conditions).
- Material density (fibre volume and resin control).

Figure 19.1 shows a sample of a specimen test ready to be used and tested by a machine.

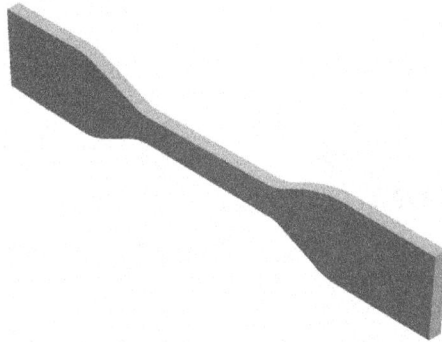

FIGURE 19.1   Specimen test.

19.1.2   DESTRUCTIVE TEST

Destructive tests are performed with the purpose of obtaining the material's mechanical properties, such as hardness, toughness, mechanical resistance, and ductility, and through this method being able to verify the quality of the material used. These are very important tests since they show in a proportional way how the materials perform in different situations.

The main destructive tests usually performed on manufactured composite material parts are as follows:

- Tensile tests
- Bending tests
- Compression tests
- Flat shear tests
- Interlaminar shear tests
- Fatigue tests
- Interlaminar fracture tests
- Impact tests, like compression tests after impact and sandwich panel tests.

19.1.3 Non-destructive Test

Non-destructive testing methods are used to detect superficial and internal imperfections in composite materials. This test is performed to ensure an acceptable quality level during the manufacturing process as well as during their working service life.

The non-destructive tests of composite materials need a very different methodology to identify defects compared to what is done with metals. Defects and imperfections could be produced during two different stages:

- Manufacturing processes
- Working service life.

19.1.3.1 Types of Non-destructive Tests Performed in Composite Materials

- Visual inspection technique
- Ultrasonic inspection
- A-scan
- C-scan
- ANDSCAN
- Thermography
- Radiography
- X-ray inspection
- Shearography
- Acoustic emission.

19.2 PROCESS CONTROL

Process control in composite materials refers to the systematic management and monitoring of manufacturing processes to ensure that the composite products meet design and quality specifications. It involves setting parameters, real-time monitoring, and making adjustments to prevent defects and optimise efficiency.

The manufacturing process and QC procedures are key elements that control the success of our work for the composite part.

The process control could include sub-procedures that will qualify, for example, the quality of the materials, control methods for an in-process procedure, and fitting qualification.

19.2.1 Parameters to Consider during the Process Control

QC of materials: Materials should be controlled following the correct inspection procedure as per the material specifications (datasheets). Considering material properties, storage requirements, curing cycles, shelf-life, etc.

Lay-up process control: The laminate schedule should be controlled to ensure the right stacking order and laminate orientation. During the lay-up process, any possible contamination should be avoided and prevented. Working environmental conditions, such as temperature and humidity, should also be kept below the specified levels.

Cure cycle control: During a cure cycle, high compaction pressure is applied to the laminate combined with high temperature; this will vary according to the manufacturing procedure. Both parameters should be controlled to prevent any internal defect formations, such as porosity or voids.

Post-processing: After the part is cured, it generally requires a post-processing procedure and/or assembly operation. However, the post-processing and assembly operations, such as trimming, drilling, and fastening, will require special care, since any machining or incorrect process could lead to heat damage or delamination in the cured laminate.

During assembly, uncontrolled handling of the composites will tend to delaminate if excessive force is used. In addition, a QC procedure during the manufacturing process of the laminates should be performed, followed by a final check. This is prior to the part being released, with non-destructive inspection (NDI/NDT).

BIBLIOGRAPHY

'Analysis and Performance of Fiber Composites' (Third edition). Bhagwan D. Agarwal, Lawrence J. Broutman and K. Chandrashekhara. Wiley India Pvt. Ltd., 2015.
'Composite Manufacturing Process Selection Using Analytical Hierarchy Process'. A. Hambali, S. M. Sapuan, Napsiah Ismail and Yusoff Nukman. *International Journal of Mechanical and Materials Engineering (IJMME)*, Vol. 4 (2009), No. 1, 49–61.

20 Thermosetting Materials during the Curing Process

Thermosetting materials are a type of polymer that, when exposed to a curing process, will undergo an irreversible chemical reaction transformation, from a liquid or semi-solid state to a rigid, cross-linked structure that gives them high mechanical strength and thermal stability. This transformation is influenced by factors such as time, temperature, and pressure.

They are commonly used in composite materials, acting as the matrix to bind the reinforcement fibres like carbon, glass, or aramid. Figure 20.1 shows a graph representing gelation and vitrification on a temperature versus time chart, where the x-axis represents gel or vitrification time on a logarithmic scale, while y-axis represents the temperature in Kelvin.

20.1 STAGES OF THE CURING CYCLE FOR THERMOSETTING MATERIALS

1- Heating and Softening

At first, the material is initially heated, causing it to soften and flow. This allows the resin to impregnate the reinforcing fibres and fill the mould.

2- Gelation

As the temperature increases, the resin will start its polymerization process where the cross-linking occurs as part of the chemical reaction, and the material will transition from a liquid to a gel state. This means the material loses flow characteristics and begins to form a network structure. This is the gel point, where the polymer becomes partially solidified but is not yet fully cured, and its time depends on temperature and the resin's reactivity. Higher temperatures will accelerate the gelation.

3- Curing

Further heating continues the cross-linking process, forming a rigid three-dimensional network. The curing temperature and time depend on the resin system, like epoxy, polyester, and phenolic, and the final size or thickness of the desired product.

Common curing techniques include autoclave curing, oven curing, and hot press curing.

4- Vitrification

Vitrification occurs when the resin's cross-linking has progressed to the point where the material reaches or exceeds its glass transition temperature (T_g) during

DOI: 10.1201/9781003565222-20

curing. It is a change of state for thermosets, transitioning from a liquid or rubber state to a glassy and rigid state as a result of an increase in molecular mass. Vitrification typically occurs at lower curing temperatures, where heat diffusion is limited, or when curing is incomplete.

5- **Post-Curing (Optional)**

Additional heating at a higher temperature ensures complete cross-linking and increases the material's properties, such as thermal resistance and dimensional stability.

Monitoring the curing of a thermosetting resin becomes difficult due to the interaction between chemical kinetics and the changes in the physical properties during the reaction. These kinetics are affected by the local viscosity, which can vary as a function of reaction conversion and temperature.

That being said, the stop of the reaction does not necessarily mean that the reaction has been completed, as it could have been only deactivated due to vitrification. If the resin temperature rises again later, we may sometimes be able to get the reaction to continue.

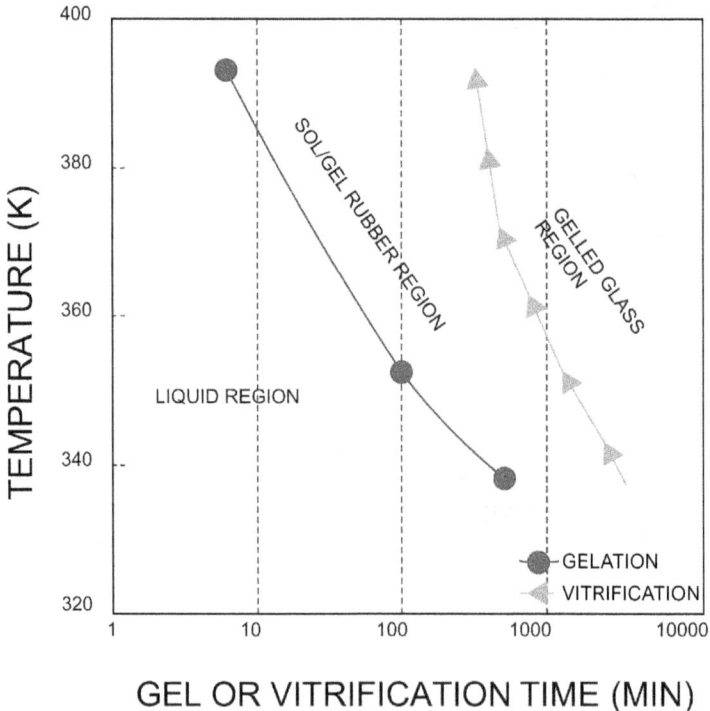

FIGURE 20.1   Gelation and vitrification representation on a temperature vs time chart.

20.2   CURING PROCESS TIMES AND INFLUENCES

20.2.1   Process Times

- Gelation Time: Depends on the resin system and cure temperature; typically ranges from a few minutes to hours.

- Complete Cure Time: Can range from 1 to 24 hours or more, depending on the resin, thickness, and curing method.
- Post-Cure Time: 21 Varies between 1 to 12 hours at elevated temperatures, depending on the desired Tg increase.

20.2.2 IMPORTANT FACTORS DURING CURING

- Temperature: Higher temperatures speed up the reaction but can cause exothermic hotspots in thick parts.
- Pressure: Reduces voids and allows for proper fibre-matrix bonding.
- Resin Composition: Reactivity and filler content affect curing speed and the degree of cross-linking.
- Thickness: Thicker parts require longer curing times to ensure uniform heat distribution and curing throughout the whole surface.

In the polymerization reaction, thermosetting polymers may be formed in two ways:

- **Polymerization of monomers:** A mixture of monomers reacts via a polymerization mechanism
- **Cross-linking:** Macromolecules previously created are chemically cross-linked. Before this, we could find the macromolecules in a linear or branched form. In either case, the polymerization reaction will depend on the concentration of the reacting components.

20.3 CONVERSION-TEMPERATURE-TRANSFORMATION (CTT)

To obtain a better understanding of the curing process, the concept of the conversion-temperature-transformation (CTT) or "Conversion vs Temperature phase diagram" for the curing of thermosets has been developed.

A conversion-temperature-transformation (CTT) graph is a tool used in the context of thermosetting resins to illustrate the curing behaviour of these materials as a function of temperature and the degree of conversion. It is particularly useful for understanding key curing transitions such as gelation and vitrification, and for optimizing the curing process.

The diagram shown in Figure 20.2 is a typical representation of the time required to reach gelation and vitrification.

The gelation, vitrification, and decomposition lines define the material state. The grey lines are trajectories leading to vitrification at the same glass transition temperature:

- Reaction induced
- Temperature and reaction induced
- Temperature induced

20.3.1 GELATION LINE

During gelation, a state transition from liquid to rubber occurs, characterized by the presence of linear or branched molecules with a low molecular weight, which

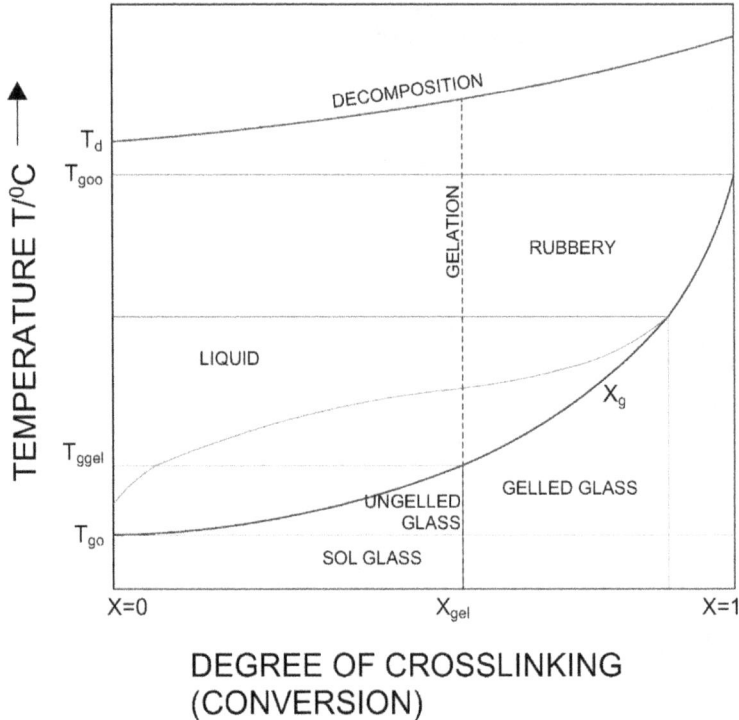

DEGREE OF CROSSLINKING (CONVERSION)

FIGURE 20.2 Typical conversion-temperature-transformation diagram (CTT).

conform a three-dimensional network providing elastic properties that the resin initially did not have.

The gelation line represents the transition from a liquid state to a gel state. Below this line, the resin remains a fluid, and crossing this line marks the gel point, where the resin cannot be processed anymore.

20.3.2 VITRIFICATION LINE

This line represents the transition from a rubbery or gel state to a glassy state, as the curing process increases the material's Tg.

Vitrification begins to occur when the glass transition temperature of the growing macromolecule is higher than the curing temperature (Tg exceeds the curing temperature), leading to the transition from the rubbery state to a gelled glass (if gelation occurs) or to an ungelled glass (if gelation does not occur).

Below the glass transition temperature of the non-cross-linked resin (Tgo), there is no reaction since any possible reactive components remain in a glassy state.

20.3.3 DECOMPOSITION LINE

There is also a degradation or decomposition area shown. If the degradation temperature (Td) overlaps with Tg∞, it will be a situation where the resin could never reach 100% conversions, as it would be fighting against its own degradation.

20.4   VISCOSITY AND ELASTIC MODULUS EVOLUTION DURING A POLYMER MOULDING CYCLE

The viscosity and elastic modulus of a thermosetting polymer matrix undergo significant changes during a polymer moulding cycle, especially for those involved in the manufacturing of composite materials. Understanding these changes is critical to ensure high-quality parts with optimal mechanical properties.

An example of such changes is shown in Figure 20.3, where after gelation, solid behaviour starts to be observed. It indicates that before gelation, the viscosity of the system can be monitored, while after gelation, the elastic modulus can be measured.

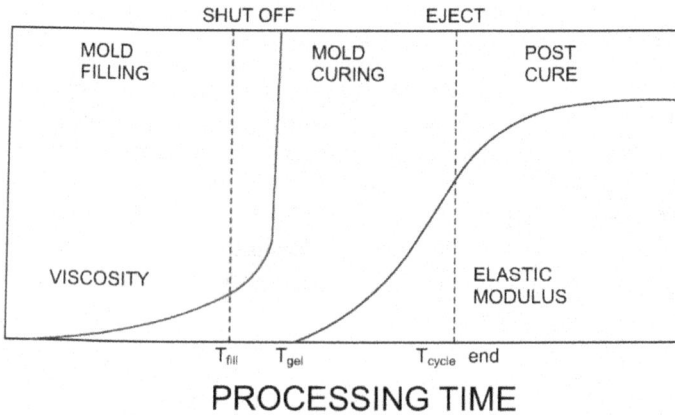

FIGURE 20.3   Evolution during polymer moulding cycle.

Viscosity: The viscosity refers to the resistance of the polymer matrix to flow. Its evolution during the moulding cycle can be separated into phases:

- Initial Heating Phase (Viscosity Decrease): As the polymer is heated, viscosity decreases due to thermal softening. This reduction facilitates the resin's flow, allowing it to wet the reinforcement fibres. For thermosetting resins, the lowest viscosity is called the minimum viscosity point.
- Gelation Phase (Viscosity Increase): Cross-linking begins as the curing reaction progresses, causing viscosity to increase rapidly.
- Final Curing Phase (Solidification): Beyond the gel point, viscosity becomes theoretically infinite as the resin solidifies into a rigid, cross-linked struc-
    This phase ensures the part achieves its final shape and strength.

        ⌐he elastic modulus (E) describes the material's stiffness. Its evo-
            ⌐sity changes during curing:

                ⌐odulus): The polymer matrix behaves like a vis-
                    ⌐ The elastic modulus is minimal because the
                        ⌐s not yet possess structural integrity.

- Gelation Phase (Rapid Modulus Increase): As cross-linking progresses, the elastic modulus increases dramatically. This reflects the material's transition from a liquid to a semi-solid, capable of supporting stress but still deformable.

- Post-Gelation to Vitrification Phase: Cross-linking continues, increasing the stiffness even further. The material enters a rubbery state, with moderate stiffness and elasticity.

- Vitrification Phase (Maximum Modulus): Vitrification occurs when the Tg (glass transition temperature) exceeds the curing temperature. The material becomes rigid and glassy, exhibiting its maximum elastic modulus.

20.5   FLOW AND CURE CYCLE

The flow and cure cycle describes the sequence of thermal, pressure, and chemical events during the processing of composite materials, particularly thermosetting polymers. This cycle involves resin flow, fibre wetting, and the chemical transformation of the resin from a liquid to a rigid, cross-linked solid.

During thermoset processing, the ability to flow stops at gelation. It is important to adapt the process to achieve the required flow prior to the gel point. This is particularly relevant if we are processing a reactive thermoset such as an adhesive or prepreg.

Process	Flow (min)	Cure (min)	Post-cure (min)
Autoclave lamination	20	150	Opt
Pultrusion	1	1	Opt
Filament winding	30	120	Opt
RTM	10	10	120

BIBLIOGRAPHY

'Analysis and Performance of Fiber Composites' (Third edition). Bhagwan D. Agarwal, Lawrence J. Broutman and K. Chandrashekhara. Wiley India Pvt. Ltd., 2015.
'Composite Manufacturing Process Selection Using Analytical Hierarchy Process'. A. Hambali, S. M. Sapuan, Napsiah Ismail and Yusoff Nukman. *International Journal of Mechanical and Materials Engineering (IJMME)*, Vol. 4 (2009), No. 1, 49–61.

21 Glass Transition Temperature

The glass transition temperature (Tg) for a cured polymeric material is a crucial thermal property, especially when the matrix material is a polymer. It can be defined as the temperature at which the material changes state as the temperature increases, from a rigid solid to a softer or semi-flexible material. At this temperature, the structure of the polymer remains intact, but the chains are no longer cross-linked.

Tg dictates the maximum service temperature for a composite. For example, composites used in aerospace applications must have matrices with high Tg to withstand elevated temperatures, while a low Tg can limit applications where thermal resistance is crucial. Tg also affects the mechanical and thermal performance of the composite, where composites operating below Tg ensure stiffness and strength, while exceeding Tg can lead to a loss of these properties. To put this with some numbers, and since the Tg is affected by humidity, the service temperature should be set below the Tg, 50°F.

Note: Do not confuse the Tg, glass transition temperature, with Gt (Gel time)

The glass transition temperature is typically represented in a temperature vs. elastic modulus graph, which, as shown in Figure 21.1, illustrates how the material stiffness changes as the temperature changes.

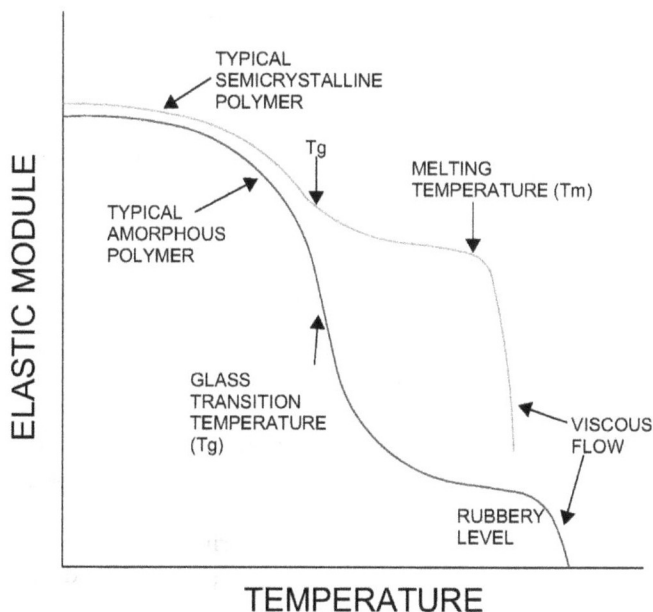

FIGURE 21.1 Temperature vs elastic modulus chart.

DOI: 10.1201/9781003565222-21

It can be observed that below Tg, the elastic modulus is high, indicating stiffness. This region is called "the glass region", where the polymer matrix is in a rigid, glassy state. As the material gets closer to the Tg, the molecular interaction in the polymer matrix increases, causing the modulus to drop and transitioning from a rigid to a softer state. Beyond Tg, the polymer matrix becomes rubbery and flexible. In this region, the elastic modulus is significantly lower, and the composite reinforcement's properties dominate.

Another way to see it is through a "cure time vs temperature" graph. In Figure 21.2, the resin is dissolved in a solvent and impregnated into a woven fabric. The first part of the graph shows the working time period, where the resin is still liquid enough to be applied and no tackiness is observed. In the next area of the graph, the temperatures are set to provide the right number of cross-linking reactions, which, as seen previously, thermosets typically start as small molecules, where the chain extension followed by cross-linking ends in a gelled network.

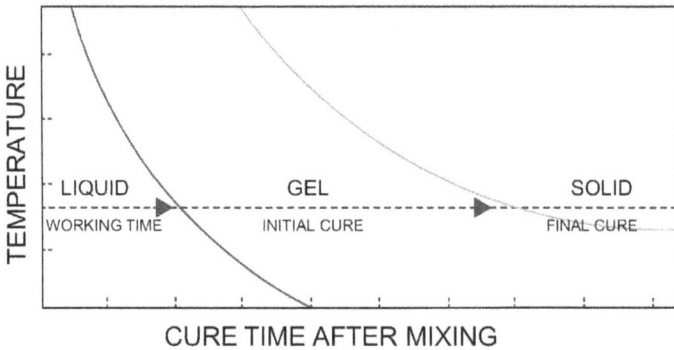

FIGURE 21.2   Cure time (after mixing) vs temperature.

The temperature area is critical since, as the higher the temperature is, the shorter the time to reach the gel curve will be. During this period, the thermoset cannot pass this gel point.

21.1   TEMPERATURE VS TIME TRANSFORMATION (TTT)

To fully understand the curing process, a very informative concept called the temperature vs time transformation diagram (TTT) for the curing of thermosets was created, as shown in Figure 21.3.

The TTT diagrams are a representation of the curing temperature of a material over a period of time, where some of the transformations that a thermoset can undergo are shown (gelling, vitrification, decomposition, etc.).

TTT diagrams are a really useful tool for monitoring the curing process or transformation of thermosetting materials and their compounds. The different areas of the TTT diagram are limited by the gelation and vitrification curves, with the particularity that the Tg will have a unique value in this gelation curve, equal to Tggel (temperature at which gelation and vitrification occur at the same time).

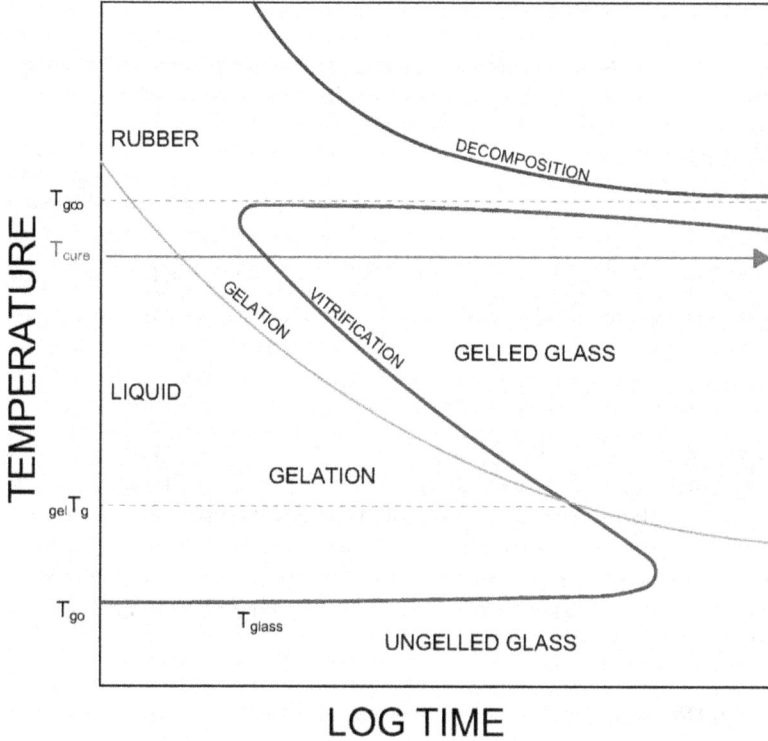

FIGURE 21.3   TTT diagram.

On the other hand, there is a Tg∞, which can be defined as the Tg of the fully cured material, or as the maximum temperature at which the material vitrifies. There is also a Tgo, which is the Tg of the material when it has not yet started to react.

In the vitrification curve, it is observed that immediately above Tgo, the vitrification time presents a maximum due to the opposite influences that they present regarding temperature, viscosity, and the rate constant of the reaction. Below Tg∞, the vitrification time passes through a minimum due to the opposite influences of the rate constant of the reaction concerning temperature (higher temperature, higher rate).

21.1.1   POINTS TO BE CONSIDERED ON THE TTT DIAGRAM

- If the material is cured at a constant temperature and above Tg, only the gelation is observed.
- If the curing temperature is between Tg gel and Tg during the reactive process, gelation and vitrification are observed.
- If the curing temperature is below Tg gel, only vitrification is observed.

21.2 STRESS VS STRAIN

At the beginning of the material selection stage process, it is important to understand and know the difference between their mechanical properties:

- Modulus: How stiff it is?
- Strength: How much stress it can support before suffering permanent deformation?
- Toughness: How much energy can be absorbed before fractures or breaks?
- Fracture Toughness: How fast the cracks propagate when exposed to fatigue under load?
- Hardness: How hard it is?

The stress vs. strain curve for composite materials reflects their mechanical behaviour when exposed to loading. This curve depends on the type of composite, the matrix and reinforcement properties, and the direction of loading relative to the reinforcement.

Stress-strain curves are an extremely important graphical measure of a material's mechanical properties and are widely used to understand the strength, deformation, and failure criteria of any material.

21.2.1 GENERATION OF A STRESS VS STRAIN CURVE

The stress-strain curve is plotted during the tensile test, where the material to be tested is exposed to a load condition until failure while a plotter records the stress and strain data from it. Figure 21.4 shows a typical example of a stress vs strain curve and its different stages.

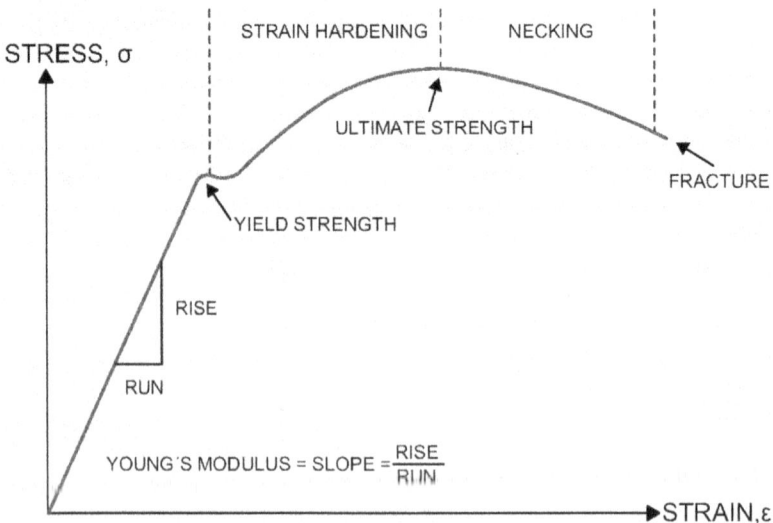

FIGURE 21.4   Typical stress vs strain curve.

- **Yield Strength:** Is the maximum stress the material experiences after going through elongation and the permanent deformation starts. Once the yield strength of a material is reached, large deformation occurs with every small increase in load. If the yield point is never reached, then the material will return to its shape once the stress is removed.

  In the stress-strain curve, yield strength is the point where the curve changes its trajectory to strain. For some materials, the yield strength in the stress-strain curve is quite noticeable, but for a few others, it is not.

- **Ultimate Tensile Strength:** The ultimate tensile strength or tensile strength of a material is the maximum stress value of the stress-strain curve. This is the maximum stress value that the material could experience before failure.

- **Young's Modulus:** Young's modulus is defined as the ratio of stress to strain. It is a measure of the stiffness of elastic materials.

21.2.2 STRESS-STRAIN CURVE FOR SOME TYPICAL FIBRES USED IN COMPOSITES

There are many typical common reinforcement forms, such as unidirectional, woven fabrics (bidirectional), non-continuous chopped fibres, and particulate fillers. All these different reinforcement forms can be placed in a stress vs strain curve as shown in Figure 21.5.

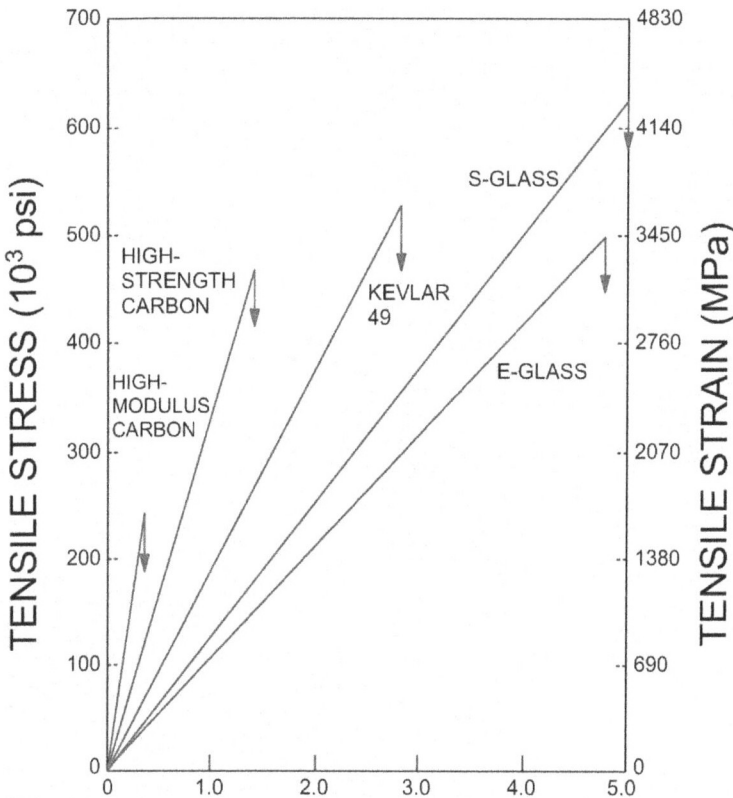

FIGURE 21.5 Stress vs strain curve for typical fibres.

The most common type of composite reinforcement is E-glass. From the stress-strain data, E-glass has the lowest modulus but has relatively good tensile strength, while S-glass offers a higher modulus and tensile strength but is more expensive compared to E-glass.

Kevlar has both a higher modulus and tensile strength if compared with the glass fibres.

Carbon fibre has the highest modulus and strength. It has about the same tensile strength as E-glass but has a considerably higher modulus.

Note that for the high modulus carbon fibre, the tensile strength decreases, so this fibre would be used where the modulus and stiffness are set as important design criteria.

BIBLIOGRAPHY

'Analysis and Performance of Fiber Composites' (Third edition). Bhagwan D. Agarwal, Lawrence J. Broutman and K. Chandrashekhara. Wiley India Pvt. Ltd., 2015.

'Analysis of Composite Materials. A Survey'. Z. Hashin. *Journal of Applied Mechanics*, Vol. 50 (1983), No. 3, 481–505.

'Composite Manufacturing Process Selection Using Analytical Hierarchy Process'. A. Hambali, S. M. Sapuan, Napsiah Ismail and Yusoff Nukman. *International Journal of Mechanical and Materials Engineering (IJMME)*, Vol. 4 (2009), No. 1, 49–61.

'Engineering Mechanics of Composite Materials' (Second edition). Isaac M. Daniel and Ori Ishai. New York: Oxford University Press, 2006.

'Handbook: An Engineering Compendium on the Manufacture and Repair of Fiber Reinforced Composites'. R. L. Ramkumar, N. M. Bhatia, J. D. Labor and J. S. Wilkes. NJ, USA: Department of Transportation FAA Technical Center: Atlantic City International Airport, 1987.

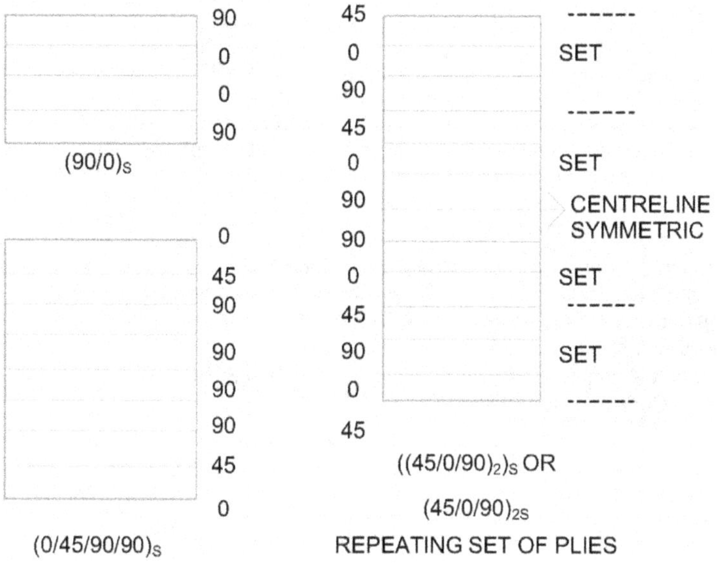

FIGURE 22.3 Symmetric laminate codification.

them also uses the subscript "2", which is helpful when a large laminate sequence needs to be written.

22.2 LAMINATE PROPERTIES

The properties of solid polymer laminates depend on the specific materials used, the number of layers, the layers' orientation, and the manufacturing process. However, to optimise their performance for specific applications, a number of properties need to be taken into consideration, such as:

- The adhesion capacity of the matrix system used to bond the fibres and the layers together.
- The fibre type used in each layer.
- The geometry or the fibre angle in each layer.
- The ratio between the matrix system and fibre reinforcement.
- The cure temperature.
- The compression pressure during the cure process.
- Good adhesion properties to the reinforcement fibres.
- Mechanical resistance to traction/compression/shear (Tensile strength – Modulus of elasticity – Elongation).
- Chemical resistance (osmosis/corrosion).
- Thermal resistance.
- Resistance to fatigue.
- UV Resistance.

BIBLIOGRAPHY

'Analysis and Performance of Fiber Composites' (Third edition). Bhagwan D. Agarwal, Lawrence J. Broutman and K. Chandrashekhara. Wiley India Pvt. Ltd., 2015.

'Composite Materials'. S. C. Sharma. Alpha Science International, Ltd, 2000.

'Composite Materials: Design and Applications'. Daniel Gay, Suong V. Hoa and Stephen W. Tsai. CRC Press LLC, 2003.

'Engineering Mechanics of Composite Materials' (Second edition). Isaac M. Daniel and Ori Ishai. New York: Oxford University Press, 2006.

'Handbook: An Engineering Compendium on the Manufacture and Repair of Fiber Reinforced Composites'. R. L. Ramkumar, N. M. Bhatia, J. D. Labor and J. S. Wilkes. NJ, USA: Department of Transportation FAA Technical Center: Atlantic City International Airport, 1987.

'Introduction to Composite Materials'. Hong T. Hahn and Stephen W. Tsai. Technomic Publishing Company, Inc., 1980.

23 Composite Systems

To understand what composite systems are, it is needed to visualise the fibre-reinforced material's properties and stresses as vectors relative to three principal axes, based on the orientation of the fibres, which are shown in Figure 23.1 as Direction 1, Direction 2, and Direction 3. This coordinate system is essential for understanding the anisotropic behaviour of composites, where material properties differ along different directions.

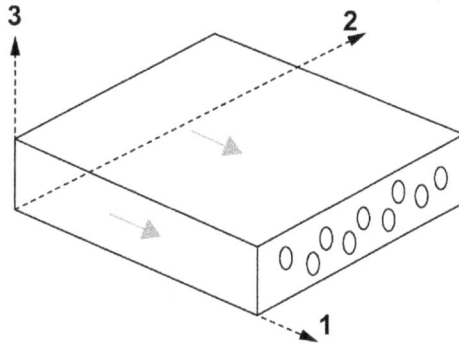

FIGURE 23.1 Composite coordinate system relative to fibres.

As mentioned, composite systems are always defined relative to the fibres, based on the following directions:

- **Direction 1 (Longitudinal direction):** The axis is parallel to the length of the fibres; this is the strongest and stiffest direction in the composite because the fibres are primarily aligned along this axis, and fibres carry most of the load in this direction.

- **Direction 2 (Transverse Direction – In-plane):** The axis is perpendicular to the fibres, aligned within the plane of the lamina or laminate; it represents the in-plane transverse direction where the matrix is in charge of carrying the load.

- **Direction 3 (Transverse Direction – Out-of-plane or normal to fibre):** The axis is perpendicular to both Direction 1 and Direction 2, typically through the thickness of the laminate. It represents the out-of-plane behaviour, which is critical for understanding interlaminar stresses and delamination.

DOI: 10.1201/9781003565222-23

23.1   COMPOSITE STRESS SYSTEM

A composite stress system refers to the stress distribution within a composite material; when composites are exposed to the application of an external load, the stress response will not be uniform since each component has different mechanical properties, such as stiffness and strength.

A stress component is defined by two factors:

1) The plane in which the stress is acting, as shown in Figure 23.2.

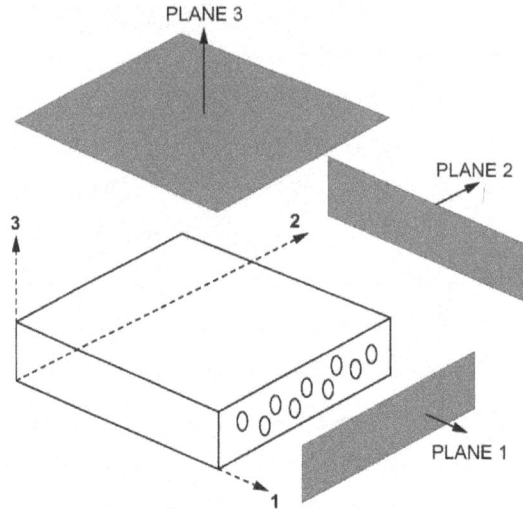

FIGURE 23.2   Planes where the stress can act.

2) The direction of the stress, as shown in Figure 23.3.

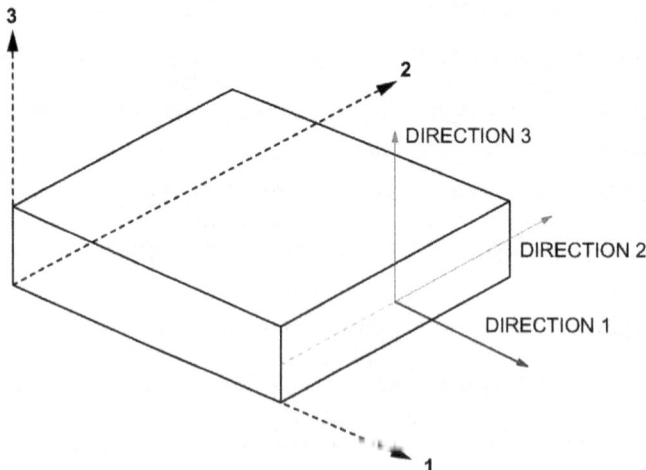

FIGURE 23.3   Representation of stress direction within a plane.

An example of the Direct and Shear stress components acting on a plane is shown in Figure 23.4; in this case, it is represented on Plane 1. It is important to remark that the planes in a composite system are usually expressed with the letter "$S_{plane,direction}$".

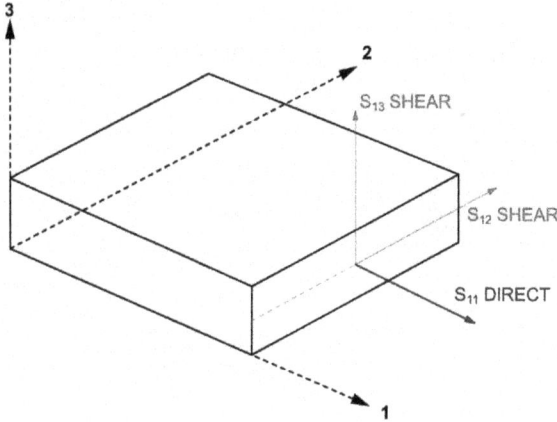

FIGURE 23.4   Representation of stress direction within a plane.

By convention, the nomenclature for **direct** and **shear** stresses is expressed in the following way:

S11 = S1
S22 = S2
S33 = S3

23.2   3D ORTHOTROPIC STRESS SYSTEM

A 3D orthotropic composite stress system involves direct stresses (normal stresses) and shear stresses, which are distributed along and between the main material directions (Direction 1, Direction 2, and Direction 3). Shear stresses act parallel to a plane, and they are defined by two directions: S12 and S13 for Plane 1. Figure 23.5 represents a direct and shear stress in a full 3D orthotropic stress system.

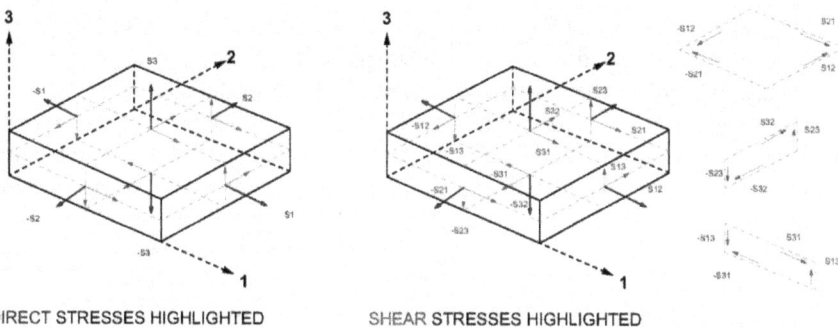

DIRECT STRESSES HIGHLIGHTED   SHEAR STRESSES HIGHLIGHTED

FIGURE 23.5   Direct and shear stress in a full 3D orthotropic stress system.

23.3   REDUCED 2D ORTHOTROPIC STRESS SYSTEM

A reduced 2D orthotropic composite stress system focuses mainly on the in-plane behaviour of a composite material, ignoring through-thickness stresses and strains. This simplification is common in the analysis of laminated composites, where the in-plane directions dominate the structural response. Figure 23.6 represents a direct and shear stress in a reduced 2D orthotropic stress system.

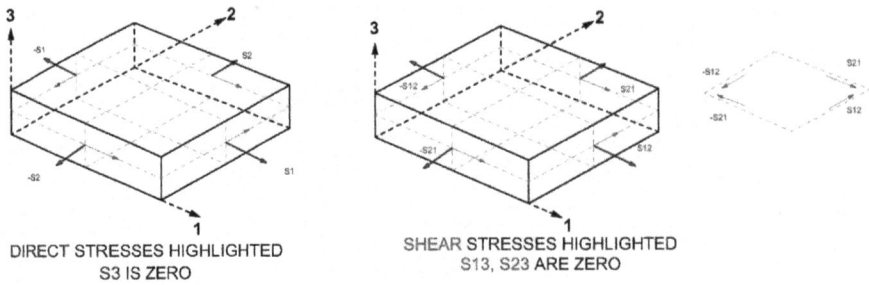

DIRECT STRESSES HIGHLIGHTED
S3 IS ZERO

SHEAR STRESSES HIGHLIGHTED
S13, S23 ARE ZERO

FIGURE 23.6   Direct and shear stress in a reduced 2D orthotropic stress system.

23.4   STIFFNESS AND STRENGTH RATIOS
IN A COMPOSITE SYSTEM

The stiffness and strength of a composite depend purely on the properties and arrangement of the matrix and the reinforcement, as well as their interaction with each other.

The stiffness describes the resistance of a material to deformation under loads. For composites, the longitudinal stiffness is directly related to and dependent on the reinforcement stiffness, while the transverse stiffness will depend on the orientation and distribution of the fibres, with lower contributions from the reinforcement compared to axial alignment.

Strength in composites could be defined as the ability of the composite to withstand maximum stress without failure. While the composite's strength can be calculated or approximated, it is less predictable than stiffness since it will depend on its failure or fracture mechanism, and this can vary.

In terms of ratios, a high stiffness ratio, which involves the relation between the fibres and the matrix modulus (E_f / E_m), is necessary in composites to maximise the stiffness. While a strength ratio, which involves the strengths of the fibre and matrix (σ_f / σ_m), ensures that the reinforcement will support most of the loads before the matrix fails.

Pure fibre is stiff and strong, but comparing them and putting them all together in a comparison chart as shown in Figure 23.7, it can be observed, for example, that glass and graphite fibres vs. steel and aluminium, the composite systems are more flexible and weaker.

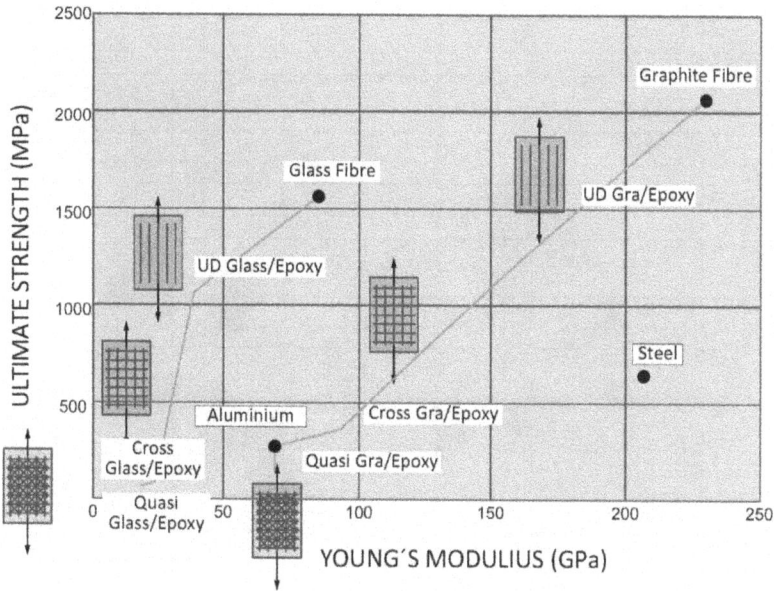

FIGURE 23.7 Stiffness and strength ratios in a composite system for different materials.

Glass Fibre/Epoxy system:

- Unidirectional (0 degrees)
- Cross Plies (0/90 degrees)
- Quasi-isotropic (0/45/–45/90/90/–45/45/0)

To continue with some comparisons and looking at the specific gravity for different materials, the following analysis can be done from the plotted graph in Figure 23.8.

- The density (or specific gravity) of steel is around four times higher than glass or graphite fibres.
- The density of aluminium is more comparable to glass or graphite fibres
- The density of fibre/resin systems is slightly less than fibres due to the presence of epoxy
- Variations in composite lay-up ply orientations do not change the density

Fibres seem to have a tremendous advantage. However, Graphite/Epoxy systems approach Aluminium and Steel

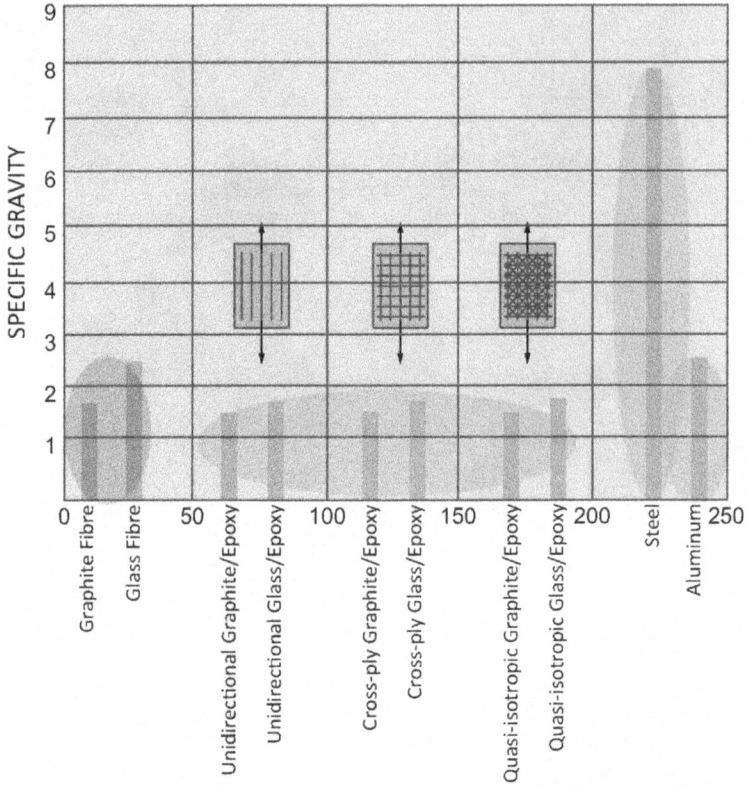

FIGURE 23.8 Specific gravity comparison.

TABLE 23.1
Material data comparison.

Material	Specific Gravity	Young's Modulus (GPa)	Ultimate Strength (MPa)	Specific Modulus (GPa-m³/Kg)	Specific Strength (GPa-m³/Kg)
Graphite fibre	1.8	230	2067	0.128	1.148
Glass fibre	2.5	85	1550	0.034	0.620
Unidirectional graphite/epoxy	1.6	181	1500	0.113	0.938
Unidirectional glass/epoxy	1.8	39	1062	0.021	0.590
Cross-ply graphite/epoxy	1.6	96	373	0.060	0.233
Cross-ply glass/epoxy	1.8	24	88	0.013	0.049
Quasi-isotropic graphite/epoxy	1.6	70	276	0.044	0.173
Quasi-isotropic glass/epoxy	1.8	19	73	0.011	0.041
Steel	7.8	207	648	0.027	0.831
Aluminium	2.6	69	276	0.027	0.100

23.4.1 Why Does a Composite Strength and Stiffness Change when It Is Made in a Unidirectional UD Fibre/Resin System?

As mentioned before, the strength and stiffness of a composite material in a unidirectional (UD) fibre/resin system is directly related to the directional alignment of the fibres and their interaction with the matrix.

So, in terms of the reinforcement direction in a UD composite, where the fibres are aligned in a single and unique direction, the orientation maximises strength and stiffness in the fibre direction (longitudinal direction) because the load is purely carried by the high-strength fibres. When aligned, the fibres support the majority of the applied load along their length due to their strong tensile strength and stiffness when compared to the matrix. Figure 23.9 shows the representation of fibres aligned with the load direction.

The opposite would happen if a transverse load is applied, as shown in Figure 23.10, where perpendicular to the fibre direction (transverse direction) is the worst

FIGURE 23.9   Representation of UD fibres aligned with load direction.

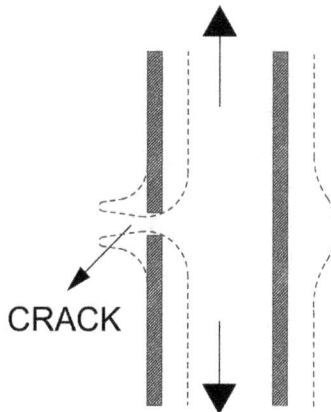

FIGURE 23.10   Representation of cracking due to transverse loading direction on a UD fibres.

scenario for the fibres. They will have a poor contribution in supporting the loads due to their weakness in shear and compression when these loads are off-axis. And here is when the matrix comes into action.

As mentioned in previous chapters, the matrix holds, supports, and keeps the fibres together, but it also plays a key role in the transverse strength and stiffness of the composites, as it provides resistance in the perpendicular direction, transferring the loads to the fibres through shear and also helping the fibres against buckling when they are exposed to compression. The matrix typically has lower strength and stiffness compared to the fibres, making the composite weaker in the transverse direction.

So, the most important thing to care about in a UD composite system is the interaction between fibres and the matrix, since cracks can propagate more easily in the transverse direction through the weaker matrix, further reducing strength and stiffness.

Here are some key points to remember:

- When fibres are perfectly isolated in a system, they can have incredibly strong and stiff properties.
- In practice, the fibres will have breaks and kinks and will not be perfectly aligned.
- They cannot be used only in fibre form; they need to incorporate the matrix into the system, such as a resin.
- The interaction between fibre and matrix is complex.

If considering a unidirectional, longitudinal loading on a fibre/resin system:

- The fibres are not perfect and may have different levels of cracking.
- The matrix is relatively weak, but it provides the linking to keep the fibres together.
- Resin properties are tested and in general considered more predictable.
- The strength /stiffness is a result of two ingredients: fibre and resin.

23.4.2 Why Does a Composite Strength and Stiffness Change when Ply Orientation Is Changed from UD?

In a UD composite, all the fibres are aligned in a single direction, making the material highly anisotropic, but in this particular case where the ply orientation changes from the UD direction, the strength and stiffness will change as well because the fibres are no longer aligned with the primary load direction.

Composite materials can carry loads in multiple directions when a multi-directional lay-up is applied. However, this redistribution of fibres across different orientations will reduce the strength and stiffness in the longitudinal direction since fewer fibres are aligned, which means fewer of them will carry the load in that direction. The stiffness in any direction will always depend on the proportion of fibres that are aligned with that single direction. Figure 23.11 represents a predicted failure load vs ply angle, where fibre orientation deviates from the loading direction. It can be appreciated how stiffness decreases when fibres start to rely on the matrix to transfer load through shear.

When plies are oriented at different angles, the stress will be transferred along the fibres and matrix, where off-axis fibres will experience shear stresses and transverse tensile/compressive stresses, which can lead to earlier failure compared to purely axial loading in a UD composite. These cross-ply orientations are more isotropic (uniform properties in different directions), and the interaction between longitudinal, transverse, and shear strengths and stiffness becomes more complex. For example, a [0°/90°] lay-up balances strength in the orthogonal directions, while a [0°/±45°/90°] lay-up, as shown in Figure 23.12, provides a balance for in-plane and shear properties, sacrificing maximum stiffness in any single direction.

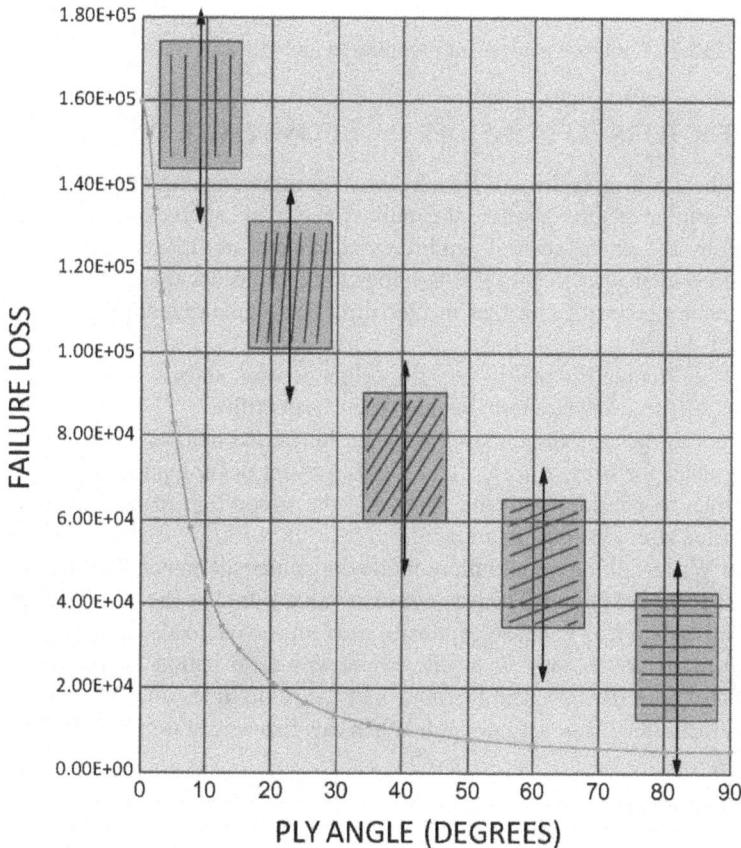

FIGURE 23.11   Predicted failure load vs ply angle.

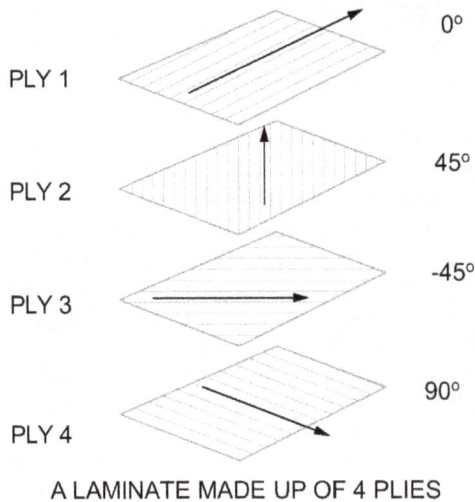

PLY 1   0°

PLY 2   45°

PLY 3   -45°

PLY 4   90°

A LAMINATE MADE UP OF 4 PLIES

FIGURE 23.12   Laminate made up of four plies in multiple directions.

23.5   FACTORS TO CONSIDER IN A COMPOSITE SYSTEM

The composite is a system in which fibres (reinforcements) coexist in a matrix (resin or similar medium), where the main strength and stiffness characteristics are provided by the fibres, such as unidirectional or woven (fibre structure can vary), followed by their directionality. It is important to consider that aligned fibres will provide higher properties for that specific direction, while random orientations yield isotropic behaviour.

The fibre volume fraction is very important as well, since a higher fibre content increases stiffness and strength, but also reduces ductility.

And not to mention the bonding between fibres and matrix, since a good adhesion is crucial for stress transfer, and it will prevent delamination or debonding. It is important to consider both the fibres and the supporting matrix in the material stiffness and strength considerations.

Note: When talking about failure modes in composite materials, there are basically three types of failure. The first one to mention would be the fibre reinforcement, which fails when it is exposed to tensile or compressive loads. If the matrix fails, this will occur under shear or tensile stress. The third option for potential failure will involve both the fibre and the matrix at the same time, since it is the interface between them that could actually fail. When this happens, it is called "debonding".

BIBLIOGRAPHY

'Analysis and Performance of Fiber Composites' (Third edition). Bhagwan D. Agarwal, Lawrence J. Broutman and K. Chandrashekhara. Wiley India Pvt. Ltd., 2013.
'Basic Mechanics of Laminated Composite Plates'. A. T. Nettles. Alabama: Marshall Space Flight Center-MSFC (NASA), 1994.
'Composite Materials'. S. C. Sharma. Alpha Science International, Ltd, 2000.

'Composite Materials: Design and Applications'. Daniel Gay, Suong V. Hoa and Stephen W. Tsai. CRC Press LLC, 2003.

'Composite Materials Handbook. Volume I – Polymer Matrix Composites Guidelines for Characterization of Structural Materials – MIL-HDBK-17'. McGraw Hill/Departments and Agencies of the Department of Defence, 2002.

'Engineering Materials 1 – An Introduction to their Properties and Applications' (Second edition). Michael F. Ashby and David R. H. Jones. UK: Department of Engineering, University of Cambridge, 1996.

'Engineering Mechanics of Composite Materials' (Second edition). Isaac M. Daniel and Ori Ishai. New York: Oxford University Press, 2006.

'Fiber Composite Analysis and Design: Composite Materials and Laminates. Volume I'. Z. Hashin, B. W. Rosen, E. A. Humphreys, C. Newton and S. Chaterjee. Washington, DC: U.S. Department of Transportation: Federal Aviation Administration Office of Aviation Research, 1997.

'Handbook: An Engineering Compendium on the Manufacture and Repair of Fiber Reinforced Composites'. R. L. Ramkumar, N. M. Bhatia, J. D. Labor and J. S. Wilkes. NJ, USA: Department of Transportation FAA Technical Center: Atlantic City International Airport, 1987.

'Introduction to Composite Materials'. Hong T. Hahn and Stephen W. Tsai. Technomic Publishing Company, Inc., 1980.

'Introduction to Composite Materials Design'. Ever J. Barbero. USA: Department of Mechanical & Aerospace Engineering – West Virginia University/Taylor & Francis, 1998.

'Laminated Composite Plates'. David Roylance. Cambridge, MA: Department of Materials Science and Engineering Massachusetts Institute of Technology, 2000.

'Marine Composites' (Second edition). Eric Green Associates. MD: Eric Green Associates Inc., 1999.

'Mechanics of Composite Materials' (Second edition). Robert M. Jones. Taylor & Francis, 1999.

'Mechanics of Laminated Composite Plates and Shells - Theory and Analysis' (Second edition). J. N. Reddy. CRC Press, 2004.

'Shigley's Mechanical Engineering Design' (Tenth edition). Richard G. Budynas and J. Keith Nisbett. McGraw Hill Education, 2015.

'The Theory of Composites'. Graeme W. Milton. Cambridge: University of Utha Press, 2002.

'Theory of Composites Design'. Stephen W. Tsai. Department of Aeronautics and Astronautics: Stanford University, 1992.

24 Laminated Composite Plates Analysis

One of the most common representations of composite materials is a cross-ply laminated structure, which is basically a stack of plies (or lay-ups) made by unidirectionally reinforced plies, as Figure 24.1 shows with a [0°/+45°/90°/−45°/0°] lay-up. As an assumption, the orientation of each ply would be arbitrary, and the lay-up sequence would be according to the final properties that are intended to be achieved.

To do this, it is recommended to know how these laminates are designed or which tools are used to analyse them.

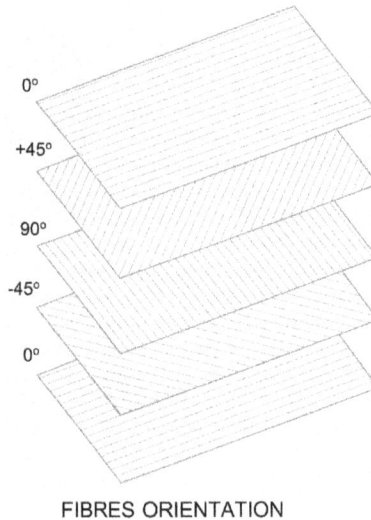

FIBRES ORIENTATION

FIGURE 24.1   Stack of cross-ply [0°/+45°/90°/−45°/0°] lay-up.

24.1   NORMAL STRESS AND STRAIN, UNIAXIALLY APPLIED LOAD

Composite materials are typically anisotropic, meaning their mechanical properties vary within their direction, and when a uniaxial load is applied (say in the x-direction), the stress and strain distribution depend on the orientation of the fibres and the stacking sequence.

24.1.1   Stress Components

For a single ply with fibres oriented at an angle θ against the load, the global stress components are related to the applied stress σ_x through transformation equations:

 DOI: 10.1201/9781003565222-24

$$\sigma_1 = \sigma_x \cos^2\theta + \sigma_y \sin^2\theta + \tau_{xy} \sin\theta \cos\theta$$

$$\sigma_2 = \sigma_x \sin^2\theta + \sigma_y \cos^2\theta - \tau_{xy} \sin\theta \cos\theta$$

$$\tau_{12} = \left(\sigma_x - \sigma_y\right)\sin\theta \cos\theta + \tau_{xy}\left(\cos^2\theta - \sin^2\theta\right)$$

where

σ_1, σ_2: stresses in the fibre and transverse directions.

τ_{12}: shear stress in the fibre coordinate system.

24.1.2 STRAIN COMPONENTS

The strain in the global x and y directions relates to the stress via the laminate's stiffness matrix:

$$\begin{bmatrix} \varepsilon_x \\ \varepsilon_y \\ \gamma_{xy} \end{bmatrix} = \begin{bmatrix} A_{11} & A_{12} & A_{16} \\ A_{12} & A_{22} & A_{26} \\ A_{16} & A_{26} & A_{66} \end{bmatrix}^{-1} \begin{bmatrix} N_X \\ N_Y \\ N_{XY} \end{bmatrix}$$

where

ε_x, ε_y, γ_{xy}: global in-plane strains.

N_X, N_Y, N_{XY}: applied in-plane loads.

A_{ij}: extensional stiffness terms from the laminate stiffness matrix.

The stress-strain relationship for the individual plies is expressed in terms of the reduced stiffness matrix Q_{ij}.

Analysing **Stress** helps identify critical areas that may experience high localized stresses and determines whether the composite can withstand the applied load without failure. As a general rule, the maximum stress in the material must not exceed the strength of the fibres, matrix, or their interface.

The **strain**, on the other hand, helps to predict deformation, which can vary significantly between longitudinal (fibre-aligned) and transverse (matrix-dominated) directions. This will ensure the material deforms within acceptable limits under load since excessive strain can lead to permanent deformation, delamination, or micro-cracking.

Putting this in simple words, **normal stress** is defined as the force per unit area acting perpendicular to that area, while the **strain** is defined as the elongation (or stretch) per unit length of a material acting in the load direction. For isotropic materials, the relation between stress and strain is independent of the load direction since only an elastic constant is needed to describe the stress-strain relation during a uniaxial applied load, and it is called "Young's modulus". For example, for an isotropic material, the stress-strain relationship is defined by the following expression:

$$\sigma = E \times \varepsilon$$

where

σ = stress

E = elastic modulus

ε = strain

For an orthotropic material, at least two elastic constants are needed to describe the stress-strain relationship in a material.

1) The mentioned Young's modulus (E_x) in one principal direction.

2) Poisson's ratio (ν_{xy}), which describes the coupling between axial and lateral strain in the same principal plane.

24.1.3 POISSON'S RATIO

Poisson's ratio is defined as the ratio in which the material changes in width per unit (transverse contraction) or in length per unit (longitudinal stretch) as a result of strain. This is represented on Figure 24.2 where stress strain is applied to a composite.

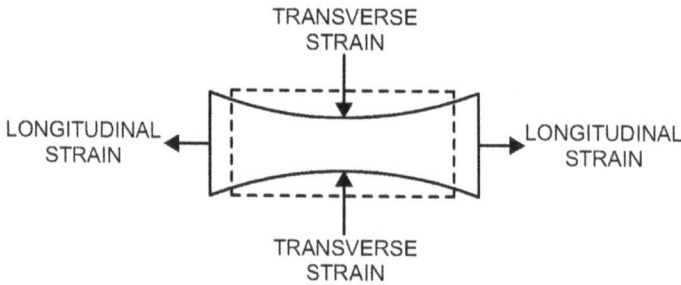

FIGURE 24.2   Representation of stress strain applied to a composite.

$$Poisson's \, Ratio \, \nu_{ij} = \frac{\varepsilon_j}{\varepsilon_i} = \frac{Transverse \, Strain}{Longitudinal \, Strain}$$

If considering a Single Ply in the principal material directions (fibre and transverse):

$$Poisson's \, Ratio \, \nu_{12} = \frac{E_2}{E_1} \nu_{21}$$

$$Poisson's \, Ratio \, \nu_{21} = \frac{E_1}{E_2} \nu_{12}$$

where

ν_{12} : Poisson's ratio for longitudinal stress, which induce transverse strain.

E_1, E_2: longitudinal and transverse modulus.

If considering an entire laminate, the effective Poisson's ratio is directly related to the laminate stiffness matrix:

$$Poisson's\ Ratio\ \nu_{e\!f\!f.} = -\frac{\varepsilon_y}{\varepsilon_x}$$

where

ε_y: strain in the transverse direction due to uniaxial stress.

ε_x: strain in the longitudinal direction.

Even under a uniaxial load, Poisson's effect induces transverse strains. Understanding this behaviour is important for

- Multiaxial Load Conditions: A uniaxially loaded composite might still experience transverse stresses due to constraints or interactions with other parts of the structure.
- Coupling Effects: Laminates with asymmetric lay-ups exhibit coupling between axial, transverse, and shear deformations, making strain analysis essential for predicting the overall behaviour.

24.2 STRESS-STRAIN RELATIONSHIP

There are different ways to express the stress-strain relationship for composite materials since they are different in the longitudinal axes (fibre-aligned) and transverse axes (perpendicular to fibres) directions due to their anisotropic properties. This relationship can be expressed in both the local or main material axes and the global coordinate systems.

24.2.1 STRESS-STRAIN RELATIONSHIP IN THE MAIN MATERIAL AXES

For a composite lamina in the longitudinal-transverse axes, the local stress-strain relationship is described as follows:

$$\begin{Bmatrix} \sigma_L \\ \sigma_T \\ \tau_{LT} \end{Bmatrix} = \begin{bmatrix} Q_{11} & Q_{12} & 0 \\ Q_{12} & Q_{22} & 0 \\ 0 & 0 & 2Q_{66} \end{bmatrix} \begin{Bmatrix} \varepsilon_L \\ \varepsilon_T \\ \frac{1}{2}\gamma_{LT} \end{Bmatrix}$$

where $\sigma_1, \sigma_2, \tau_{12}$ are the normal and shear stresses in the fibre-aligned coordinate system, while the $\varepsilon_1, \varepsilon_2, \gamma_{12}$ are the strains in the fibre-aligned system.

24.2.2 STRESS-STRAIN RELATIONSHIP IN THE GLOBAL AXES

When the lamina is oriented at an angle θ to the global coordinate system, as shown in Figure 24.3, the stress-strain relationship must be transformed using the coordinate transformation matrix.

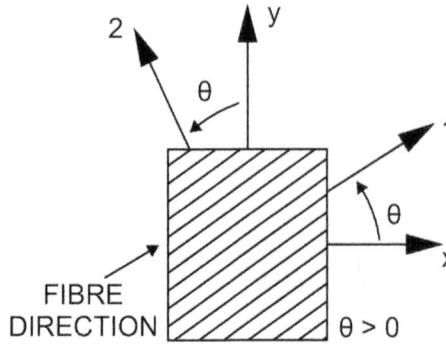

FIGURE 24.3 Representation of a lamina oriented at an angle θ to the global coordinate system.

The global stress-strain components σ_x, σ_y, τ_{xy} and ε_x, ε_y, γ_{xy} are related to the local stress-strain σ_1, σ_2, τ_{12} and ε_1, ε_2, γ_{12}. And the stress-strain relationship in the global axes is derived using the local stiffness matrix Q, which is transformed to the global stiffness matrix.

$$\begin{Bmatrix} \sigma_x \\ \sigma_y \\ \tau_{xy} \end{Bmatrix} = [T]^{-1} \begin{bmatrix} Q_{11} & Q_{12} & 0 \\ Q_{12} & Q_{22} & 0 \\ 0 & 0 & 2Q_{66} \end{bmatrix} [T] \begin{Bmatrix} \varepsilon_x \\ \varepsilon_y \\ \dfrac{1}{2}\gamma_{xy} \end{Bmatrix}$$

where the coordinate transformation matrix is expressed as follows:

$$[T]^{-1} \begin{bmatrix} \cos^2\theta & \sin^2\theta & -2\sin\theta\cos\theta \\ \sin^2\theta & \cos^2\theta & 2\sin\theta\cos\theta \\ \sin\theta\cos\theta & -\sin\theta\cos\theta & \cos^2\theta - \sin^2\theta \end{bmatrix}$$

For laminated plates, the stress-strain relationship is defined by the global stiffness matrix (A) for in-plane loads:

$$\begin{Bmatrix} N_x \\ N_y \\ N_{xy} \end{Bmatrix} = \begin{bmatrix} A_{11} & A_{12} & A_{16} \\ A_{12} & A_{22} & A_{26} \\ A_{16} & A_{26} & A_{66} \end{bmatrix} \begin{Bmatrix} \varepsilon_x \\ \varepsilon_y \\ \gamma_{xy} \end{Bmatrix}$$

where

A_{ij}: depend on the properties of individual plies and their orientations.

$$A_{ij} = \sum_{k=1}^{n} Q_{ij}^{(2)} \left(z_k - z_{k-1} \right)$$

And where $Q_{ij}^{(2)}$ are the transformed stiffness terms for the k-ply thickness, and z_k, z_{k-1} are the top and bottom positions of the k-ply thickness.

For a single ply, however, the stress-strain relationship is defined as follow:

$$\begin{bmatrix} \sigma_1 \\ \sigma_2 \\ \tau_{12} \end{bmatrix} = \begin{bmatrix} Q_{11} & Q_{12} & Q_{16} \\ Q_{12} & Q_{22} & Q_{26} \\ Q_{16} & Q_{26} & Q_{66} \end{bmatrix} \begin{bmatrix} \varepsilon_1 \\ \varepsilon_2 \\ \gamma_{12} \end{bmatrix}$$

where Q_{ij} are terms of the stiffness matrix in the local fibre coordinate system.

24.3 STRESS-STRAIN RELATIONSHIP

The stress-strain relationship in composite materials significantly differs between isotropic and orthotropic plates due to the directional dependence related to material properties in orthotropic plates.

24.3.1 ISOTROPIC PLATE

An isotropic material has uniform properties in all directions, as represented in Figure 24.4 where no particular alignment is shown, and the stress-strain relationship for isotropic materials under plane stress would be expressed as follows:

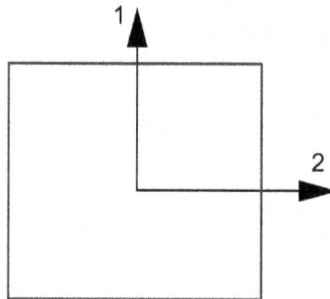

FIGURE 24.4   Isotropic plate

$$
\begin{bmatrix} \varepsilon_x \\ \varepsilon_y \\ \gamma_{xy} \end{bmatrix} = \begin{bmatrix} \dfrac{1}{E} & -\dfrac{\nu}{E} & 0 \\ -\dfrac{\nu}{E} & \dfrac{1}{E} & 0 \\ 0 & 0 & \dfrac{1}{G} \end{bmatrix} \begin{bmatrix} \sigma_x \\ \sigma_y \\ \tau_{xy} \end{bmatrix}
$$

where

- E: Young's modulus (same in all directions).
- ν: strain in the longitudinal direction.
- $G = \dfrac{E}{2(1+\nu)}$: shear modulus (derived from E and ν)

Stiffness in 1-direction = Stiffness in 2-direction = Stiffness in 3-direction.

24.3.2 ORTHOTROPIC PLATE

An orthotropic material has unique mechanical properties along each of the three perpendicular axes. In the case of Figure 24.5, fibres are aligned with the direction 1 of the perpendicular axes, where, the stress-strain relationship for an orthotropic lamina would be expressed as follows:

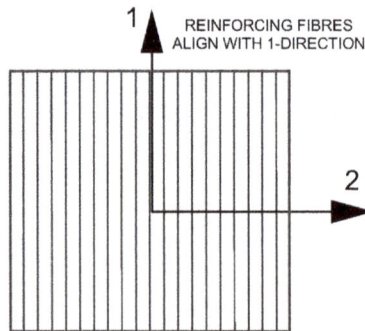

FIGURE 24.5 Orthotropic plate.

$$
\begin{bmatrix} \varepsilon_x \\ \varepsilon_y \\ \gamma_{xy} \end{bmatrix} = \begin{bmatrix} \dfrac{1}{E_x} & -\dfrac{\nu_{xy}}{E_x} & 0 \\ -\dfrac{\nu_{yx}}{E_y} & \dfrac{1}{E_y} & 0 \\ 0 & 0 & \dfrac{1}{G_{xy}} \end{bmatrix} \begin{bmatrix} \sigma_x \\ \sigma_y \\ \tau_{xy} \end{bmatrix}
$$

where
- E_x, E_y: Young's modulus in x and y directions.
- ν_{xy}, ν_{yx}: Poisson's ratios for coupling between x and y axes, related by symmetry of

$$\frac{\nu_{xy}}{E_x} = \frac{\nu_{yx}}{E_y}$$

- G_{xy}: shear modulus in the x-y plane.

Stiffness in 1-direction >>Stiffness in 2-direction ≠ Stiffness in other directions.

24.4 LAMINATE ANALYSIS IN-PLANE

In-plane stresses σ_x, σ_y, τ_{xy} are expressed and led by the previously mentioned relationship between stress and strain, modified for isotropic and orthotropic materials.

In a 2D isotropic plate, material properties are uniform in all directions within the plane, and the stress-strain relationships in 2D are derived using plane stress or plane strain assumptions. This means that an isotropic material can be defined by any two of the three properties presented in the relationship below:

$$G = \frac{E}{2(1+\nu)}$$

In 3D, isotropic materials have the same properties in all directions; however, for 3D orthotropic plates, material properties differ along three principal axes (x, y, z), and the properties below need to be considered:

E_1 = Young's modulus along the longitudinal axis of the fibre.
E_2 = Young's modulus transverse to the fibre.
E_3 = Young's modulus through thickness.
G_{12} = in-plane shear modulus.
G_{23} = through-thickness 23 shear modulus.
G_{31} = through-thickness 31 shear modulus
ν_{12} = primary Poisson's ratio
ν_{23} = through-thickness 23 Poisson's ratio
ν_{31} = through-thickness 31 Poisson's ratio.

For the orthotropic system, the direction must be specified. For example,

$$\sigma_1 = E_1 \ x \ \varepsilon_1 \ \text{ or } \sigma_2 = E_2 \ x \ \varepsilon_2$$

But, in general, laminate plates will experience stresses in more than one direction. If the applied load acts either parallel or perpendicular to the fibres, then the plate will be considered fully orthotropic.

There is also an important factor to be considered in this analysis, and that is Poisson's ratio. Different expressions are used depending on whether the load is applied along the fibres or if it is transverse to the fibres.

$$Poisson's\ Ratio\left(for\ loading\ along\ the\ fibres\right)\nu_{12} = \frac{\varepsilon_T}{\varepsilon_L} = \frac{\varepsilon_2}{\varepsilon_1}$$

or

$$Poisson's\ Ratio\left(for\ loading\ transverse\ to\ the\ fibres\right)\nu_{12} = \frac{\varepsilon_L}{\varepsilon_T} = \frac{\varepsilon_1}{\varepsilon_2}$$

Consider a single ply loaded in the fibre longitudinal direction and free for contraction in the transverse direction, as shown in Figure 24.6; as a result, the transverse stress and through thickness stress are zero:

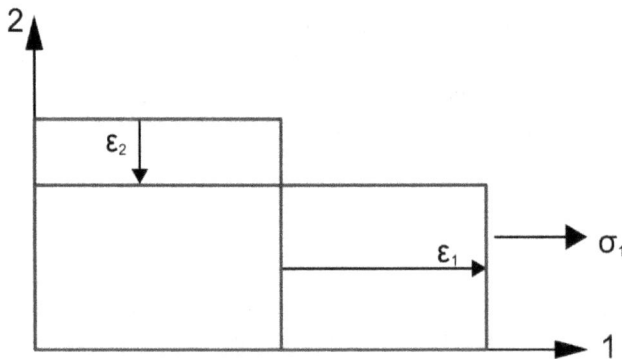

FIGURE 24.6 Ply loaded in the fibre longitudinal direction and free for contraction in the transverse direction.

$$\sigma_1 = 0$$
$$\sigma_3 = 0$$

$$\varepsilon_1 = \frac{\sigma_1}{E_1}$$

$$\varepsilon_2 = -\nu_{12}\,\varepsilon_1\ \text{then}\ \varepsilon_2 = -\nu_{12}\,\frac{\sigma_1}{E_1}$$

$$\varepsilon_3 = -\nu_{13}\,\varepsilon_1\ \text{then}\ \varepsilon_3 = -\nu_{13}\,\frac{\sigma_1}{F_1}$$

Now consider the ply loaded in the fibre transverse direction and free for deformation in the longitudinal direction, as shown in Figure 24.7; as a result, the longitudinal stress and through-thickness stress is zero:

$$\sigma_1 = 0$$
$$\sigma_3 = 0$$

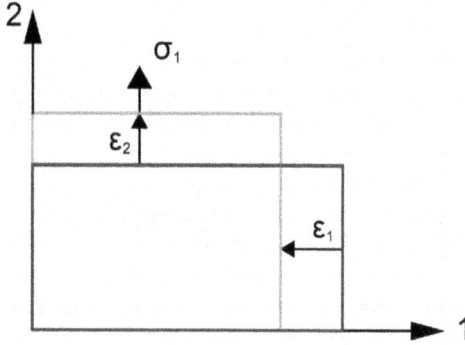

FIGURE 24.7 Ply loaded in the fibre transverse direction and free for deformation in the longitudinal direction.

$$\varepsilon_2 = \frac{\sigma_2}{E_2}$$

$$\varepsilon_1 = -\nu_{21}\varepsilon_2 \text{ then } \varepsilon_1 = -\nu_{21}\frac{\sigma_2}{E_2}$$

$$\varepsilon_3 = -\nu_{23}\varepsilon_2 \text{ then } \varepsilon_3 = -\nu_{23}\frac{\sigma_2}{E_2}$$

Now consider the ply loaded in the through thickness direction and free to contract in the other directions; as a result, the longitudinal and transverse fibre stress are zero:

$$\varepsilon_3 = \frac{\sigma_3}{E_3}$$

$$\varepsilon_1 = -\nu_{31}\varepsilon_3 \text{ then } \varepsilon_1 = -\nu_{31}\frac{\sigma_3}{E_3}$$

$$\varepsilon_2 = -\nu_{32}\varepsilon_3 \text{ then } \varepsilon_2 = -\nu_{32}\frac{\sigma_3}{E_3}$$

If we consider the direct strains applied with all direct stresses simultaneously, as represented in Figure 24.8, it will show that the longitudinal fibre strain will be:

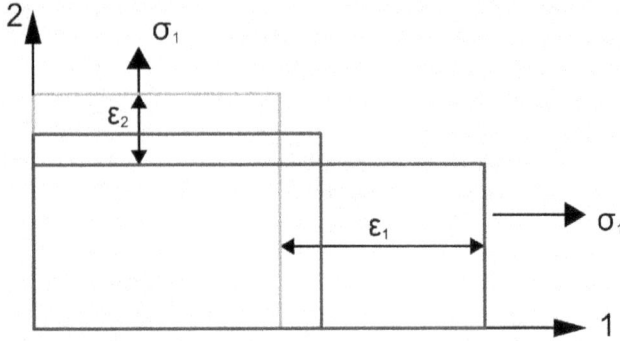

FIGURE 24.8 Direct strain applied in all directions at the same time.

$$\varepsilon_1 = \frac{\sigma_1}{E_1} - \nu_{21}\,\varepsilon_2 - \nu_{31}\,\varepsilon_3 = \frac{\sigma_1}{E_1} - \nu_{21}\frac{\sigma_2}{E_2} - \nu_{31}\frac{\sigma_3}{E_3}$$

Similarly to this, the transverse fibre strain will be expressed as follows:

$$\varepsilon_2 = \frac{\sigma_2}{E_2} - \nu_{12}\,\varepsilon_1 - \nu_{32}\,\varepsilon_3 = \frac{\sigma_2}{E_2} - \nu_{12}\frac{\sigma_1}{E_1} - \nu_{32}\frac{\sigma_3}{E_3}$$

And at last, the through-thickness strain will be expressed as follows:

$$\varepsilon_3 = \frac{\sigma_3}{E_3} - \nu_{13}\,\varepsilon_1 - \nu_{23}\,\varepsilon_2 = \frac{\sigma_3}{E_3} - \nu_{13}\frac{\sigma_1}{E_1} - \nu_{23}\frac{\sigma_2}{E_2}$$

These formulas express the strain components stretched due to an applied load, minus the contraction of Poisson's effect due to another load perpendicular to this applied force.

Shear forces will also play a role here; shear stress and shear strain are related by a constant, like the normal stresses and strains. This constant is called the shear modulus and is usually denoted by the letter G:

The in-plane shear and through-thickness shears shown in Figure 24.9 are related by the following expression:

$$\tau_{12} = G_{12}\,\gamma_{12}$$

$$\tau_{23} = G_{23}\,\gamma_{23}$$

$$\tau_{31} = G_{31}\,\gamma_{31}$$

where τ_{12}, for example, is the shear stress (1 and 2 indicate shear in the 1–2 plane), and γ_{12} is the shear strain

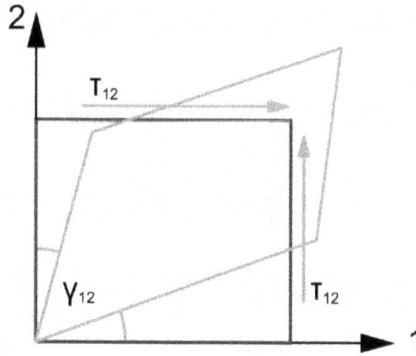

FIGURE 24.9 In-plane shear and through-thickness shears.

All these equations can be written in a matrix form named **compliance matrix.**

24.5 COMPLIANCE MATRIX

The compliance matrix is used to relate stresses and strains in anisotropic materials, and it is particularly useful for analysing the mechanical response of composite laminates. It can be defined as the inverse of the stiffness matrix and is denoted as $[S]$.

Assuming plane stress as $\sigma_3 = \tau_{31} = \tau_{23} = 0$

So, the relationship between stress and strain in linear elastic materials can be written as follows:

$$\{\varepsilon\} = [S]\{\sigma\}$$

where
- $\{\varepsilon\}$ is the strain vector.
- $[S]$ is the compliance matrix.
- $\{\sigma\}$ is the stress vector.

For a three-dimensional stress analysis, the compliance matrix $[S]$ will be expressed as a 6x6 symmetric matrix:

$$
\begin{bmatrix} \varepsilon_x \\ \varepsilon_y \\ \varepsilon_z \\ \gamma_{xy} \\ \gamma_{yz} \\ \gamma_{zx} \end{bmatrix} =
\begin{bmatrix}
S_{11} & S_{12} & S_{13} & 0 & 0 & 0 \\
S_{12} & S_{22} & S_{23} & 0 & 0 & 0 \\
S_{13} & S_{23} & S_{33} & 0 & 0 & 0 \\
0 & 0 & 0 & S_{44} & 0 & 0 \\
0 & 0 & 0 & 0 & S_{55} & 0 \\
0 & 0 & 0 & 0 & 0 & S_{66}
\end{bmatrix}
\begin{bmatrix} \sigma_x \\ \sigma_y \\ \sigma_z \\ \tau_{xy} \\ \tau_{yz} \\ \tau_{zx} \end{bmatrix}
$$

For orthotropic materials, the compliance coefficients will depend on the material properties.

$$E_x, E_y, E_z = \text{Young's modulus along the main material direction.}$$

$$\nu_{xy}, \nu_{yz}, \nu_{zx} = \text{Poisson's ratios.}$$

$$G_{xy}, G_{yz}, G_{zx} = \text{shear modulus.}$$

And the compliance matrix coefficients are follows:

$$S_{11} = \frac{1}{E_x}$$

$$S_{22} = \frac{1}{E_y}$$

$$S_{33} = \frac{1}{E_z}$$

$$S_{12} = -\frac{\nu_{xy}}{E_x}$$

$$S_{13} = -\frac{\nu_{xz}}{E_x}$$

$$S_{23} = \frac{\nu_{yz}}{E_y}$$

$$S_{44} = \frac{1}{G_{yz}}$$

$$S_{55} = \frac{1}{G_{zx}}$$

$$S_{66} = \frac{1}{G_{xy}}$$

For 2D plane stress conditions, the compliance matrix can be reduced to a 3x3 matrix form:

$$\begin{bmatrix} \varepsilon_x \\ \varepsilon_y \\ \gamma_{xy} \end{bmatrix} = \begin{bmatrix} S_{11} & S_{12} & 0 \\ S_{12} & S_{22} & 0 \\ 0 & 0 & S_{66} \end{bmatrix} \begin{bmatrix} \sigma_x \\ \sigma_y \\ \tau_{xy} \end{bmatrix}$$

where

$$S_{11} = \frac{1}{E_x}$$

$$S_{22} = \frac{1}{E_y}$$

$$S_{12} = -\frac{\nu_{xy}}{E_x} = -\frac{\nu_{yx}}{E_y}$$

$$S_{66} = \frac{1}{G_{xy}}$$

By inverting the compliance matrix, we can get stress as a function of strain, obtaining the **reduced stiffness matrix**

24.6 REDUCED STIFFNESS MATRIX

For orthotropic materials, both matrices, the compliance and the reduced matrices, are symmetric and invertible as long as the material has a non-zero modulus.

$$\begin{bmatrix} \sigma_x \\ \sigma_y \\ \tau_{xy} \end{bmatrix} = \begin{bmatrix} Q_{11} & Q_{12} & Q_{16} \\ Q_{12} & Q_{22} & Q_{26} \\ Q_{16} & Q_{26} & Q_{66} \end{bmatrix} \begin{bmatrix} \varepsilon_x \\ \varepsilon_y \\ \gamma_{xy} \end{bmatrix}$$

where:

$Q_{66} = G_{xy}$ in-plane shear modulus.

$$Q_{11} = \frac{E_x}{1 - \nu_{xy}\nu_{yx}}$$

$$Q_{22} = \frac{E_y}{1 - \nu_{xy}\nu_{yx}}$$

$$Q_{12} = \frac{\nu_{xy}E_y}{1 - \nu_{xy}\nu_{yx}} = \frac{\nu_{yx}E_x}{1 - \nu_{xy}\nu_{yx}}$$

The Q's are referred to the reduced stiffnesses and the matrix is abbreviated as $[Q]$. So, this inverted matrix could be named the "reduced stiffness matrix $[Q]$"

BIBLIOGRAPHY

'Analysis and Performance of Fiber Composites' (Third edition). Bhagwan D. Agarwal, Lawrence J. Broutman and K. Chandrashekhara. Wiley India Pvt. Ltd., 2015.
'Basic Mechanics of Laminated Composite Plates'. A. T. Nettles. Alabama: Marshall Space Flight Center-MSFC (NASA), 1994.
'Composite Materials'. S. C. Sharma. Alpha Science International, Ltd, 2000.

'Composite Materials: Design and Applications'. Daniel Gay, Suong V. Hoa and Stephen W. Tsai. CRC Press LLC, 2003.

'Composite Materials Handbook. Volume I – Polymer Matrix Composites Guidelines for Characterization of Structural Materials – MIL-HDBK-17'. McGraw Hill/Departments and Agencies of the Department of Defence, 2002.

'Engineering Materials 1 – An Introduction to their Properties and Applications' (Second edition). Michael F. Ashby and David R. H. Jones. UK: Department of Engineering, University of Cambridge, 1996.

'Engineering Mechanics of Composite Materials' (Second edition). Isaac M. Daniel and Ori Ishai. New York: Oxford University Press, 2006.

'Fiber Composite Analysis and Design: Composite Materials and Laminates. Volume I'. Z. Hashin, B. W. Rosen, E. A. Humphreys, C. Newton and S. Chaterjee. Washington, DC: U.S. Department of Transportation: Federal Aviation Administration Office of Aviation Research, 1997.

'Handbook: An Engineering Compendium on the Manufacture and Repair of Fiber Reinforced Composites'. R. L. Ramkumar, N. M. Bhatia, J. D. Labor and J. S. Wilkes. NJ, USA: Department of Transportation FAA Technical Center: Atlantic City International Airport, 1987.

'Introduction to Composite Materials'. Hong T. Hahn and Stephen W. Tsai. Technomic Publishing Company, Inc., 1980.

'Introduction to Composite Materials Design'. Ever J. Barbero. USA: Department of Mechanical & Aerospace Engineering – West Virginia University/Taylor & Francis, 1998.

'Laminated Composite Plates'. David Roylance. Cambridge, MA: Department of Materials Science and Engineering Massachusetts Institute of Technology, 2000.

'Marine Composites' (Second edition). Eric Green Associates. MD: Eric Green Associates Inc., 1999.

'Mechanics of Composite Materials' (Second edition). Robert M. Jones. Taylor & Francis, 1999.

'Mechanics of Laminated Composite Plates and Shells - Theory and Analysis' (Second edition). J. N. Reddy. CRC Press, 2004.

'Shigley's Mechanical Engineering Design' (Tenth edition). Richard G. Budynas and J. Keith Nisbett. McGraw Hill Education, 2015.

'The Theory of Composites'. Graeme W. Milton. Cambridge: University of Utha Press, 2002.

'Theory of Composites Design'. Stephen W. Tsai. Department of Aeronautics and Astronautics: Stanford University, 1992.

25 Classical Laminate Theory (CLT)

Classical laminate theory (CLT) is an analytical method created to design, evaluate, and analyse composite laminate structures. It is a way to predict how composite laminates will behave when they are exposed to external loads.

CLT is a variation or an extension of the classical plate theory, but in this case related to composite materials made by multiple plies bonded together, with the particularity that each ply could have different mechanical properties, different orientations, and is typically orthotropic. These relationships are based on a plate under in-plane loads, such as shear and axial forces, bending and twisting moments, and the following assumptions need to be made:

- Each lamina is orthotropic.
- Each lamina is homogeneous.
- Deformations are small, which means during deformation, a straight and perpendicular line to the middle surface will remain straight and perpendicular ($\gamma_{xz} = \gamma_{yz} = 0$).
- The laminate is thin and is loaded in plane stress ($\sigma_z = \tau_{xz} = \tau_{yz} = 0$).
- Each lamina is elastic, and linear elasticity is assumed.
- Plies are perfectly bonded.

25.1 EXTERNAL LOADS IN "CLASSICAL LAMINATE THEORY"

Assuming the stress on each ply will vary through the laminate thickness, the stresses must be considered as equivalent forces acting along the middle plane of the lamina, and these equivalent forces will be noted as N_i, where the subscript (i) refers to the direction of the stress. Obtaining as can be appreciated in Figure 25.1, along with the three stress resultants N_X, N_Y, and N_{XY}.

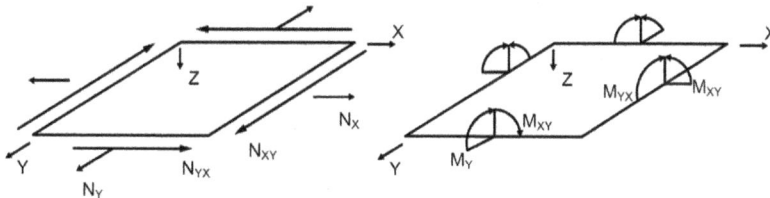

FIGURE 25.1 Resultant in-plane loads and moments on a laminate

DOI: 10.1201/9781003565222-25

Now, each stress is also producing a moment about this mid-plane, causing bending or twisting of the laminate. The moment arm will be measured at the distance z from the middle plane, resulting in M_x, M_y, and M_{xy}, all with units of torque per unit length.

where:
N_x = normal load resultant in the x direction (per unit length)
N_y = normal load resultant in the y direction (per unit length)
N_{XY} = shear load resultant (per unit length)
M_x = bending moment resultant in the yz plane (per unit length)
M_y = bending moment resultant in the xz plane (per unit length)
M_{xy} = twisting moment resultant (per unit length).

25.2   EXTERNAL LOADS IN "CLASSICAL LAMINATE THEORY"

Once the desired material/s are selected, and the four elastic moduli E_1, E_2, ν_{12}, and G_{12} are known, the values for the reduced stiffness matrix [Q] (mentioned in previous chapters) for each ply can be calculated.

$$\begin{bmatrix} \sigma_1 \\ \sigma_2 \\ \tau_{12} \end{bmatrix} = \begin{bmatrix} Q_{11} & Q_{12} & 0 \\ Q_{12} & Q_{22} & 0 \\ 0 & 0 & Q_{66} \end{bmatrix} \begin{bmatrix} \varepsilon_1 \\ \varepsilon_2 \\ \gamma_{12} \end{bmatrix}$$

where

$$Q_{66} = G_{12}$$

$$Q_{11} = \frac{E_1}{1 - \nu_{12}\nu_{21}}$$

$$Q_{22} = \frac{E_2}{1 - \nu_{12}\nu_{21}}$$

$$Q_{12} = \frac{\nu_{12}E_2}{1 - \nu_{12}\nu_{21}} = \frac{\nu_{21}E_1}{1 - \nu_{12}\nu_{21}}$$

Using the reduced stiffness matrix [Q] and applying the angle of each ply, the **transformed reduced stiffness matrix [Q] or [QB]** can be calculated as follows:

$$Q_{11} = Q_{11}\cos^4\theta + Q_{22}\sin^4\theta + 2(Q_{12} + 2Q_{66})\sin^2\theta\cos^2\theta$$

$$Q_{22} = Q_{11}\sin^4\theta + Q_{22}\cos^4\theta + 2(Q_{12} + 2Q_{66})\sin^2\theta\cos^2\theta$$

$$Q_{12} = (Q_{11} + Q_{22} - 4Q_{66})\sin^2\theta\cos^2\theta + Q_{12}(\cos^4\theta + \sin^4\theta)$$

$$Q_{66} = (Q_{11} + Q_{22} - 2Q_{12} - 2Q_{66})\sin^2\theta\cos^2\theta + Q_{66}(\sin^4\theta + \cos^4\theta)$$

$$Q_{16} = \left(Q_{11} - Q_{12} - 2Q_{66} \right) \cos^3\theta \, \sin\,\theta - \left(Q_{22} - Q_{12} - 2Q_{66} \right) \cos\,\theta + \sin^3\theta$$

$$Q_{26} = \left(Q_{11} - Q_{12} - 2Q_{66} \right) \cos\,\theta + \sin^3\theta - \left(Q_{22} - Q_{12} - 2Q_{66} \right) \cos^3\theta \, \sin\,\theta$$

The next step is to find Z_{Bottom} and Z_{Top} coordinates from the centre of the laminate or middle plane, as shown in Figure 25.2.

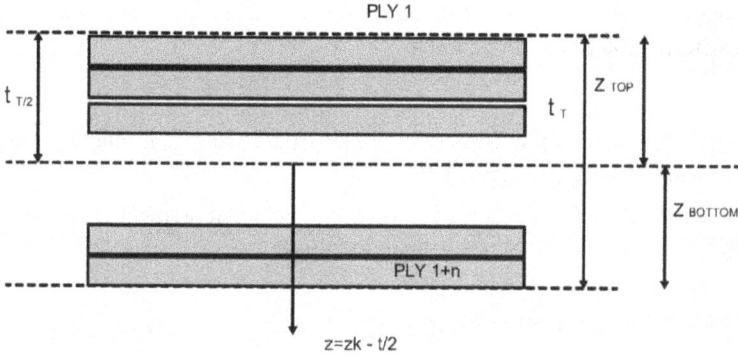

FIGURE 25.2 Laminate cross-section.

In this step, it has to be considered that the middle surface strains and curvatures (ε_0 and k's) are not included in the equation; this is due to z=0, since as was mentioned previously, these values are in the middle plane as a first assumption. The laminate stiffness matrix and the Z_k terms can be combined to new equations as follows:

$$\begin{bmatrix} \{N\} \\ \{M\} \end{bmatrix} = \begin{bmatrix} [A] & [B] \\ [B] & [D] \end{bmatrix} \begin{bmatrix} \{\varepsilon_0\} \\ \{k\} \end{bmatrix}$$

$$A_{ij} = \sum_{k=1}^{N} (Q_{ij})_k \left(z_k - z_{k-1} \right) = \sum_{k=1}^{N} (Q_{ij})_k \, t_k$$

$$B_{ij} = \frac{1}{2} \sum_{k=1}^{N} (Q_{ij})_k \left(z_k^2 - z_{k-1}^2 \right) = \sum_{k=1}^{N} (Q_{ij})_k \, t_k \, z_k$$

$$D_{ij} = \frac{1}{3} \sum_{k=1}^{N} (Q_{ij})_k \left(z_k^3 - z_{k-1}^3 \right) = \sum_{k=1}^{N} (Q_{ij})_k \left[t_k \, z_k^{-2} + \frac{t_k^3}{12} \right]$$

where:

$$\{N\} = \begin{bmatrix} N_X \\ N_Y \\ N_{XY} \end{bmatrix} : \text{in-plane loads resultant}$$

$$\{M\}=\begin{bmatrix} M_X \\ M_Y \\ M_{XY} \end{bmatrix} : \text{moment resultant}$$

$\{\varepsilon_0\}=$ mid-plane strains

$\{k\}=$ curvatures

And using the $[QB]$ matrix and the location of the Z laminate coordinates, $[A]$, $[B]$ and $[D]$ stiffness matrices can be found.

where

$[A]$ = extensional stiffness matrix. Relating the resultant in-plane loads to the in-plane strains (tensile properties).

$[B]$ = coupling stiffness matrix. Coupling the load and moment terms to the mid-plane strains and curvatures (shear coupling).

$[D]$ = bending stiffness matrix. Relating the resultant bending moments to the plate curvatures (plate bending).

By replacing the stiffness matrix values and applying the external loads, a below equation will be conformed:

$$\begin{bmatrix} N_X \\ N_Y \\ N_{XY} \\ M_X \\ M_Y \\ M_{XY} \end{bmatrix} = \begin{bmatrix} A_{11} & A_{12} & A_{16} & B_{11} & B_{12} & B_{16} \\ A_{12} & A_{22} & A_{26} & B_{12} & B_{22} & B_{26} \\ A_{16} & A_{26} & A_{66} & B_{13} & B_{26} & B_{66} \\ B_{11} & B_{12} & B_{16} & D_{11} & D_{12} & D_{16} \\ B_{12} & B_{22} & B_{26} & D_{12} & D_{22} & D_{26} \\ B_{13} & B_{26} & B_{66} & D_{13} & D_{26} & D_{66} \end{bmatrix} \begin{bmatrix} \varepsilon_{0x} \\ \varepsilon_{0y} \\ \gamma_{0XY} \\ k_x \\ k_y \\ k_{xy} \end{bmatrix}$$

Solving the six simultaneous equations will lead to finding the mid-plane strains and curvatures, which are the total laminate in-plane and bending strains.

To find the global strains on each ply, the following equations are provided:

$$\begin{bmatrix} \varepsilon_x \\ \varepsilon_y \\ \gamma_{xy} \end{bmatrix} = \begin{bmatrix} \varepsilon_{0x} \\ \varepsilon_{0y} \\ \gamma_{0XY} \end{bmatrix} \begin{bmatrix} k_x \\ k_y \\ k_{xy} \end{bmatrix}$$

Using the stress-strain equation, the global stresses can be found:

$$\begin{bmatrix} \sigma_x \\ \sigma_y \\ \tau_{xy} \end{bmatrix} = \begin{bmatrix} Q_{11} & Q_{12} & Q_{16} \\ Q_{12} & Q_{22} & Q_{26} \\ Q_{16} & Q_{26} & Q_{66} \end{bmatrix} \begin{bmatrix} \varepsilon_x \\ \varepsilon_y \\ \gamma_{xy} \end{bmatrix}$$

Then, finding the global strains, the local strains and stresses on each ply can be calculated using the transformation equation:

$$\begin{bmatrix} \sigma_x \\ \sigma_y \\ \tau_{xy} \end{bmatrix} = [T]^{-1} \begin{bmatrix} \sigma_1 \\ \sigma_2 \\ \tau_{12} \end{bmatrix}$$

$$[T]^{-1} \begin{bmatrix} \cos^2\theta & \sin^2\theta & -2\sin\theta\cos\theta \\ \sin^2\theta & \cos^2\theta & 2\sin\theta\cos\theta \\ \sin\theta\cos\theta & -\sin\theta\cos\theta & \cos^2\theta - \sin^2\theta \end{bmatrix}$$

$$\begin{bmatrix} \varepsilon_1 \\ \varepsilon_2 \\ \gamma_{12} \end{bmatrix} = [R][T][R]^{-1} \begin{bmatrix} \varepsilon_x \\ \varepsilon_y \\ \gamma_{xy} \end{bmatrix}$$

$$[T] \begin{bmatrix} \cos^2\theta & \sin^2\theta & 2\sin\theta\cos\theta \\ \sin^2\theta & \cos^2\theta & -2\sin\theta\cos\theta \\ -\sin\theta\cos\theta & \sin\theta\cos\theta & \cos^2\theta - \sin^2\theta \end{bmatrix}$$

$$[R] = \begin{bmatrix} 1 & 0 & 0 \\ 0 & 1 & 0 \\ 0 & 0 & 2 \end{bmatrix}$$

For symmetric laminates, where the ply configuration above and below the mid-plane will be mirrored between each other, the mid-plane itself acts as the neutral plane of the laminate. Consequently, the $[B]$ matrix will consist entirely of zeros.

In the case of non-symmetric laminates; where, for instance, the plies near the bottom of the plate are significantly stiffer in the x-direction, the mid-plane no longer matches with the neutral plane. Instead, the neutral plane will be shifted closer to the bottom of the laminate.

Allowing the $[B]$ matrix to contain non-zero elements will produce a mid-plane strain led by the bending strain (plate curvature). In a similar way, mid-plane strain can generate a bending moment.

Since for non-symmetric laminates, the presence of non-zero B_{ij} terms complicates the calculation of in-plane engineering constants, the same fundamental procedure as for symmetric laminates will be followed. The main difference will be in having six equations instead of three, where matrix notations simplify the process significantly.

To find ε_x, for example, only in-plane load in the x-direction needs to be considered and apply a relation between N_x and ε_{0x}, the equation will be expressed as follows:

$$
\begin{bmatrix} N_X \\ 0 \\ 0 \\ 0 \\ 0 \\ 0 \end{bmatrix} = \begin{bmatrix} A_{11} & A_{12} & A_{16} & B_{11} & B_{12} & B_{16} \\ A_{12} & A_{22} & A_{26} & B_{12} & B_{22} & B_{26} \\ A_{16} & A_{26} & A_{66} & B_{13} & B_{26} & B_{66} \\ B_{11} & B_{12} & B_{16} & D_{11} & D_{12} & D_{16} \\ B_{12} & B_{22} & B_{26} & D_{12} & D_{22} & D_{26} \\ B_{13} & B_{26} & B_{66} & D_{13} & D_{26} & D_{66} \end{bmatrix} \begin{bmatrix} \varepsilon_{0_x} \\ \varepsilon_{0_y} \\ \gamma_{0_{XY}} \\ k_x \\ k_y \\ k_{xy} \end{bmatrix}
$$

Solving the system of linear equations or inverting the ABBD matrix, curvatures and strain values will be obtained, depending on which direction the load is applied.

$$
N_X = A_{11}\,\varepsilon_{0x} + A_{12}\,\varepsilon_{0y} + A_{16}\,\gamma_{0XY} + B_{11}\,k_{0x} + B_{12}\,k_{0y} + B_{16}\,k_{0XY}
$$

BIBLIOGRAPHY

'Analysis and Performance of Fiber Composites' (Third edition). Bhagwan D. Agarwal, Lawrence J. Broutman and K. Chandrashekhara. Wiley India Pvt. Ltd., 2015.

'Basic Mechanics of Laminated Composite Plates'. A. T. Nettles. Alabama: Marshall Space Flight Center-MSFC (NASA), 1994.

'Composite Materials'. S. C. Sharma. Alpha Science International, Ltd, 2000.

'Composite Materials: Design and Applications'. Daniel Gay, Suong V. Hoa and Stephen W. Tsai. CRC Press LLC, 2003.

'Composite Materials Handbook. Volume I – Polymer Matrix Composites Guidelines for Characterization of Structural Materials – MIL-HDBK-17'. McGraw Hill/Departments and Agencies of the Department of Defence, 2002.

'Engineering Materials 1 – An Introduction to their Properties and Applications' (Second edition). Michael F. Ashby and David R. H. Jones. UK: Department of Engineering, University of Cambridge, 1996.

'Engineering Mechanics of Composite Materials' (Second edition). Isaac M. Daniel and Ori Ishai. New York: Oxford University Press, 2006.

'Fiber Composite Analysis and Design: Composite Materials and Laminates. Volume I'. Z. Hashin, B. W. Rosen, E. A. Humphreys, C. Newton and S. Chaterjee. Washington, DC: U.S. Department of Transportation: Federal Aviation Administration Office of Aviation Research, 1997.

'Handbook: An Engineering Compendium on the Manufacture and Repair of Fiber Reinforced Composites'. R. L. Ramkumar, N. M. Bhatia, J. D. Labor and J. S. Wilkes. NJ, USA: Department of Transportation FAA Technical Center: Atlantic City International Airport, 1987.

'Introduction to Composite Materials'. Hong T. Hahn and Stephen W. Tsai. Technomic Publishing Company, Inc., 1980.

'Introduction to Composite Materials Design'. Ever J. Barbero. USA: Department of Mechanical & Aerospace Engineering – West Virginia University/Taylor & Francis, 1998.

'Laminated Composite Plates'. David Roylance. Cambridge, MA: Department of Materials Science and Engineering Massachusetts Institute of Technology, 2000.

'Marine Composites' (Second edition). Eric Green Associates. MD: Eric Green Associates Inc., 1999.

'Mechanics of Composite Materials' (Second edition). Robert M. Jones. Taylor & Francis, 1999.

'Mechanics of Laminated Composite Plates and Shells - Theory and Analysis' (Second edition). J. N. Reddy. CRC Press, 2004.

'The Theory of Composites'. Graeme W. Milton. Cambridge: University of Utha Press, 2002.

'Theory of Composites Design'. Stephen W. Tsai. Department of Aeronautics and Astronautics: Stanford University, 1992.

26 Laminate

To describe what a laminate is, it could be said that a laminate is a composite made up of several stacked plies. Figure 26.1 shows a perfect example of a stack of nine plies placed one on top of the other with an alternation in their orientation, running between 0, 90, 45, and −45 degrees. To conform a laminate, there is no need to use the same material on every single ply; this laminate, for example, could be conformed even by different materials and with different orientations. When this happens, it is called a "hybrid laminate".

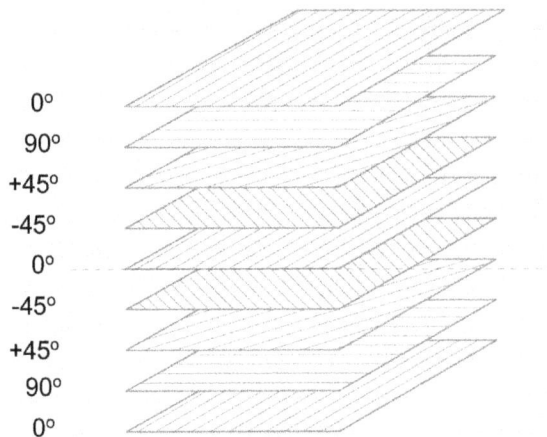

0°
90°
+45°
-45°
0°
-45°
+45°
90°
0°

FIGURE 26.1 Sample of a stack of laminate.

To describe a laminate, it is necessary to define the following characteristics of each ply that forms this laminate:

- Material: which is the fibre and the matrix to use.
- Orientation: The angle of the fibres in each ply relative to a reference axis.
- Thickness: The thickness of each ply, which is typically uniform but may vary in some designs.

In this case, a coordinate system of XYZ will be defined, instead of the 123 directions that are normally used for each ply.

 DOI: 10.1201/9781003565222-26

26.1 "LAMINA" OR "PLY"

For theoretical and calculation purposes, a lamina has been defined as a thin single layer of composite material. A "lamina" or a "ply" could be a typical composite sheet with no more than 1 mm thickness.

To materialize what a "lamina" is, it could be a stack of unidirectional or woven fibre layers contained in a matrix. In general, these unidirectional plies will have a high stiffness and strength only in the direction of the fibres, and for this reason, a coordinate system must be set to identify each of the axes.

26.2 LAMINATE ANGLE

The laminate angle refers to the fibre orientation on each lamina or ply within a laminate. The most common orientations include 0, 90, +45, −45 degrees and, in some cases, +30 and −30 degrees.

To explain this and put it in a coordinate system, Figure 26.2 is a fair representation of the referred orientations, including the different axes, such as

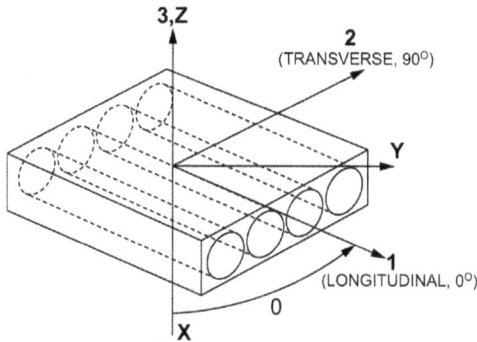

FIGURE 26.2 Representation of a ply angle in a coordinate system.

- **Axis 1** represents the direction of the unidirectional fibres in a ply. In the case of a fabric, axis 1 would be referred to as the warp.
- **Axis 2** is perpendicular to the fibre direction, contained in the sheet plane.
- The **Axis 3** will always be normal to the sheet plane. On a curved surface, the axis 3 would be normal to every single point on the surface.

In a laminate, a coordinate axis system XYZ is defined, where the laminate angle will be set by the angle between the fibre ply direction (1-2-3) and the x-axis from our coordinate system.

As a simple observation, it is also important to notice that the laminate order will be ruled by the ply sequence in each defined direction, where the angle could vary between −90 and 90 degrees.

Note: Keep in mind that a ply at 91 degrees would be the same as a ply at −89 degrees.

26.3   LAMINATE PROPERTIES: MULTIPLE LAMINA OR PLIES

The properties of a laminate will depend on the properties and orientation of the constituent materials or plies used to build it. These properties are based on the material stiffness (elastic modulus), strength, thermal expansion, density, total thickness, and stress and strain related to Poisson's ratio.

The density will be related to the fibre and matrix volume fraction values and their own densities, while the strength of a laminate may differ in the longitudinal, transverse, and shear directions. Also, composite laminates may exhibit anisotropic thermal expansion due to differing properties of fibres and the matrix.

The laminates are often described by an orientation code, for example the notation for Figure 26.3 would be:

$$\left(0° \; / −45° / 90° / +45° / 0° / 0° / +45° / 90° / −45° / 0°\right)$$

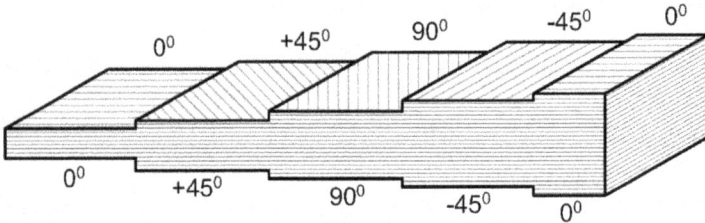

FIGURE 26.3   Stack of laminate, staggered from first ply to the last.

There is also a "shorthand" to use to describe the laminate in case it is needed, for example:

$$\left(0° \; / −45° / 90° / +45°\right)_S$$

where subscript "S" indicates symmetry.

$$\left(0° \; / −45° / 90° / +45°\right)_N$$

where subscript "N" indicates a repeated number of plies.

Other examples are as follows:

$$\left(0° \; / +45° / −45° / 90°\right)_{2S}$$

With a total number of 16 plies, due to symmetry and multiplied by two:

$$\left(0°_2 \,/+45°\,/-45°\,/90°_2\right)_S$$

With a total number of 12 plies, multiplying the last 90-degree ply by two and then a full symmetry is applied.

This laminate nomenclature describes the orientations of layers and generally assumes all layers are the same material. However, shorthands are not applicable for "hybrid laminates", for example, materials with different fibre/thickness.

BIBLIOGRAPHY

'Analysis and Performance of Fiber Composites' (Third edition). Bhagwan D. Agarwal, Lawrence J. Broutman and K. Chandrashekhara. Wiley India Pvt. Ltd., 2015.

'Analysis of Composite Materials. A Survey'. Z. Hashin. *Journal of Applied Mechanics*, Vol. 50 (1983), No. 3, 481–505.

'Basic Mechanics of Laminated Composite Plates'. A. T. Nettles. Alabama: Marshall Space Flight Center-MSFC (NASA), 1994.

'Composite Materials'. S. C. Sharma. Alpha Science International, Ltd, 2000.

'Composite Materials Handbook. Volume I – Polymer Matrix Composites Guidelines for Characterization of Structural Materials – MIL-HDBK-17'. McGraw Hill/Departments and Agencies of the Department of Defence, 2002.

'Engineering Materials 1 – An Introduction to their Properties and Applications' (Second edition). Michael F. Ashby and David R. H. Jones. UK: Department of Engineering, University of Cambridge, 1996.

'Engineering Mechanics of Composite Materials' (Second edition). Isaac M. Daniel and Ori Ishai. New York: Oxford University Press, 2006.

'Fiber Composite Analysis and Design: Composite Materials and Laminates. Volume I'. Z. Hashin, B. W. Rosen, E. A. Humphreys, C. Newton and S. Chaterjee. Washington, DC: U.S. Department of Transportation: Federal Aviation Administration Office of Aviation Research, 1997.

'Handbook: An Engineering Compendium on the Manufacture and Repair of Fiber Reinforced Composites'. R. L. Ramkumar, N. M. Bhatia, J. D. Labor and J. S. Wilkes. NJ, USA: Department of Transportation FAA Technical Center: Atlantic City International Airport, 1987.

'Handbook of Composites' (Second edition). S. T. Peters. Mountain View, CA, USA: Process Research, 1997.

'Laminated Composite Plates'. David Roylance. Cambridge, MA: Department of Materials Science and Engineering Massachusetts Institute of Technology, 2000.

'Marine Composites' (Second edition). Eric Green Associates. MD: Eric Green Associates Inc., 1999.

'Mechanics of Composite Materials' (Second edition). Robert M. Jones. Taylor & Francis, 1999.

'Mechanics of Laminated Composite Plates and Shells - Theory and Analysis' (Second edition). J. N. Reddy. CRC Press, 2004.

'The Theory of Composites'. Graeme W. Milton. Cambridge: University of Utha Press, 2002.

'Theory of Composites Design'. Stephen W. Tsai. Department of Aeronautics and Astronautics: Stanford University, 1992.

27 Introduction to Failure Criteria

When it comes to composite failure, understanding failure criteria is essential for designing, testing, and analysing composite structures. Regular testing, proper quality control during manufacturing, and knowing or predicting the environmental effects and loading conditions can help to mitigate failure risks for better optimisation.

To provide a definition of what failure is, the first step is to assume that as soon as a composite strength value is exceeded by the stress due to loading conditions, it means the structure has FAILED, or at least is no longer in conditions for further service. This is what is called "First Ply Failure Mode".

However, establishing the strength of a composite material is a task still pending. Despite the isotropic materials, the strength depends directly on the directional properties of an individual ply, which can vary along the longitudinal axis, transverse, and in shear. In addition to this, it will be different between tension and compression.

27.1 TYPICAL VARIATIONS OF FAILURE MODES

Failure in composites could happen in several modes, often depending on the type of load, material structure, and environmental conditions. Common failure types include:

- Fibre Breakage: Due to excessive tensile or compressive stress along the fibre direction. Figure 27.1 shows a typical fibre failure mode.
- Matrix Cracking: Shear or transverse tensile/compressive stresses will normally promote matrix cracking. Figure 27.2 shows a typical matrix failure mode.
- Delamination: Described as the separation of layers within a laminate, often due to interlaminar shear stresses or an impact.
- Debonding: Failure between the fibre and the matrix; this is due to poor fibre-to-matrix bonding, affecting load transfer directly.
- Buckling: Often observed under compressive loads, particularly in thin laminates.

DOI: 10.1201/9781003565222-27

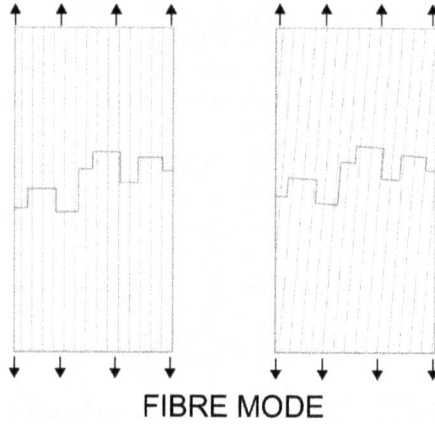

FIBRE MODE

FIGURE 27.1 Fibre failure mode.

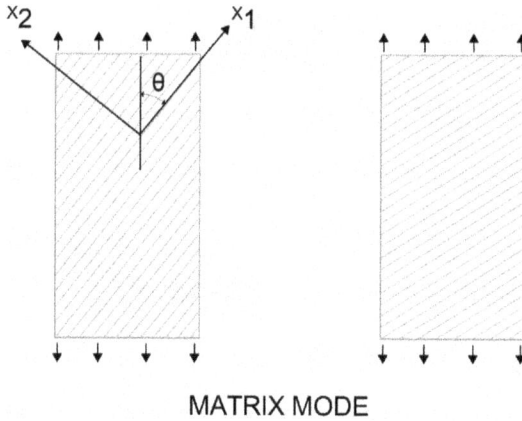

MATRIX MODE

FIGURE 27.2 Matrix failure mode.

In terms of identifying failure modes, a simple rule could help to find the one related to a laminate. Up to about 15 degrees, the failure is primarily governed by the fibre. Beyond that, the failure is fully related to the matrix.

27.2 FAILURE MECHANISMS OF COMPOSITES

The failure mechanisms of composite materials are expressed under different loading conditions. For a single ply, there are three fundamental strengths: axial, transverse, and shear; each type of strength refers to the material's resistance to failure in a specific direction relative to the fibres and the matrix. Figure 27.3, 27.4, and 27.5 represent the three different loading conditions with the referred direction.

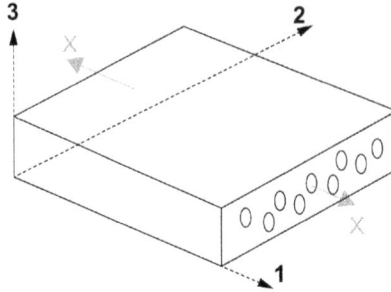

FIGURE 27.3   Axial strength in the X direction.

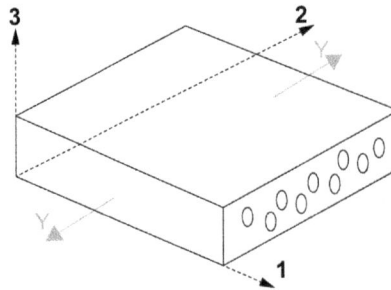

FIGURE 27.4   Transverse strength in the Y direction.

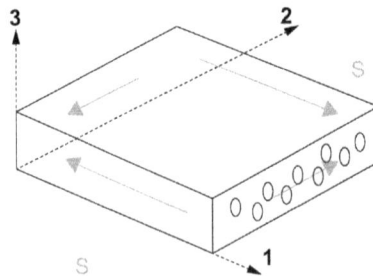

FIGURE 27.5   Shear strength in the S direction.

X = axial strength.

Y = transverse strength.

S = shear strength.

As a side note, most composites will have unequal strength in tension and compression.

X_t = axial strength (tension)

X_c = axial strength (compression)

Y_t = transverse strength (tension)

Y_c = transverse strength (compression)

S = shear strength.

27.2.1 AXIAL STRENGTH

Axial strength refers to the ability of the composite to resist an applied loading condition along the direction of the fibres. This type of failure mechanism could be in the form of tensile failure (as shown in Figure 27.6) or compressive failure, and in some cases, debonding may occur between fibres and matrix when a high load is applied and the interaction between both is in poor conditions.

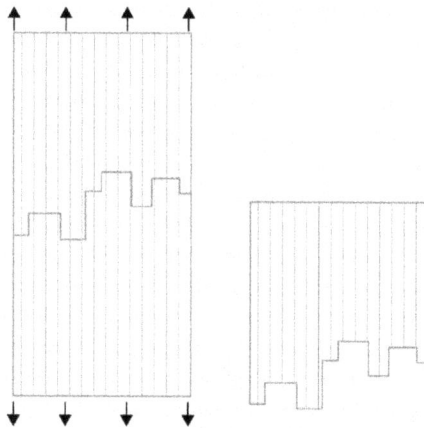

FIGURE 27.6   Representation of fibre failure under tension in the fibre direction.

Speaking of tensile failure, it is the most intuitive ply failure mode when tension is applied to the fibre direction. When the axial tensile stress exceeds the tensile strength of the fibres, the material will fail, breaking the fibres due to brittle fractures. In some cases, matrix cracking may precede the fibre failure, leading to stress concentration at the fibre ends.

Looking at it on a microscopic level, a lot is happening here:

- Fibre pulls out.
- Fibre breaking.
- Matrix cracking.
- Matrix is also trying to stabilize the fibres.
- Matrix will provide a bridging mechanism for fibre gaps.

On the other hand, for compressive failure, fibres may buckle or kink due to matrix conditions when axial load is applied. This happens especially when the matrix cannot provide sufficient lateral support to fibres anymore, accelerating the failure.

27.2.2 TRANSVERSE STRENGTH

Transverse strength refers to the composite's resistance to withstand stress applied perpendicular to the fibre direction. This strength is typically much lower than axial strength because, in this case, it is the matrix that supports it. Figure 27.7 shows a representation of fibre failure under transverse load, leading into a catastrophic crack.

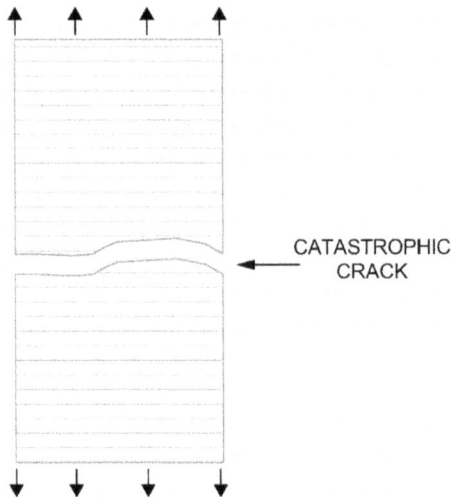

FIGURE 27.7   Representation of fibre failure under transverse load.

This type of failure could present as matrix cracking, debonding between fibre and matrix, and delamination.

For the transverse tension failure, it presents very complex behaviour on the microscopic levels:

- The crack is fully led by the matrix.
- Fibres act as stress raisers in the matrix.
- Tensile strength is less than matrix alone.

27.2.3 SHEAR STRENGTH

Shear strength describes the ability of the composite to resist shear stresses, whether in-plane (parallel to fibres, as shown in Figure 27.8) or interlaminar (between plies in laminated composites).

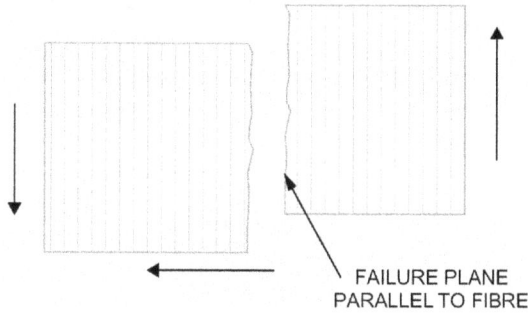

FAILURE PLANE
PARALLEL TO FIBRE

FIGURE 27.8   Representation of fibre failure under shear.

When a matrix shear failure happens, it means the shear stress exceeded the shear strength of the matrix, since the matrix is primarily responsible for resisting shear forces. Also, shear loading can cause slippage or separation between the fibre and matrix interface, reducing shear load transfer and leading to failure.

The failure mode in-plane shear (S_{12})can be assumed as a shear line failure along the matrix. At a microscopic level, the behaviour is more complex:

• Local matrix cracking behaviour leads to total failure.
• Fibres act as local stress raisers.

27.2.4 INTERLAMINAR STRENGTH (THROUGH THE THICKNESS)

As an extra mechanism of failure, interlaminar strength refers to the resistance of a composite to support stresses acting perpendicular to the laminate plane, including tensile, compressive, and shear stresses in the through-the-thickness direction.

The first failure mode under this mechanism is delamination, especially when exposed to interlaminar stress. This can be related to a defect in the bonding interface or voids that can propagate under stress. Matrix cracking is the next possible scenario, where the layer can start to separate due to matrix cracking when exposed to through-thickness stress. And at last, sliding failures could occur if interlaminar shear forces produce stress concentrations, which will lead to interlaminar shear failure.

27.3 TYPICAL DEFECTS SHOWN IN COMPOSITE MATERIALS

Defects in composites can appear during the manufacturing process, handling, or in service, and they will significantly influence their mechanical performance.

Composite materials, once cured, may have laminate defects caused by different factors such as excess or lack of resin, incorrect temperature during the curing stage, poor lamination process, etc.

The different types and most common defects in composite materials are as follows:

27.3.1 DELAMINATION

This defect occurs when two laminate sheets are pulled apart from each other. This type of separation defect can be classified as a simple defect, which could have multiple separations at different depths.

In the following image Figure 27.9, an example of a lack of union between the successive layers is presented. This produces a hole that will cause a variation in the properties of the part.

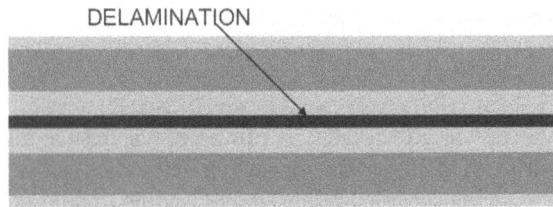

FIGURE 27.9 Delamination.

These types of defects usually occur due to a poor manufacturing process or when they are exposed to impact during handling or stacking of the laminates.

27.3.1.1 Factors That May Have Influence on Delamination during Process

- The use of different resins or catalysts in two consecutive layers.
- Insufficient impregnation of the fibres.
- Use of consecutive layers with a high fibre content.
- Lamination over a cured surface.

27.3.2 INCLUSIONS

This defect is caused by the presence of an external material that remains in the part between layers or plies during the manufacturing process. In the following image in Figure 27.10, the inclusion of a foreign object is shown in a darker tone between two sheets of laminate fabric.

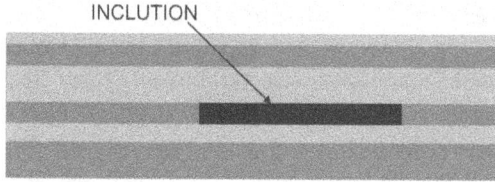

FIGURE 27.10 Inclusions.

27.3.3 VOIDS AND POROSITY

Voids and porosity in a laminate occur when the material retains air internally and has not been removed during the curing process. Multiple pores are represented in Figure 27.11 through the different layers of the laminate. These factors will lead to stress concentration.

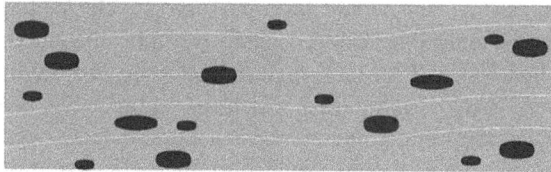

FIGURE 27.11 Porosity.

A potential cause for this defect is the lack of pressure during the manufacturing process.

Porosity can be identified in four types:

1) Uniformly distributed: The pores appear in different layers of the laminate and are produced by poor compaction between those layers.

2) Porosity in a single layer: This is produced by the accumulation of small discontinuities between two consecutive layers when the curing process has already begun, and there is not enough resin left to fill that space.

3) Interlaminar porosity: This defect appears in the form of aligned pores in the same layer due to contamination problems.

4) Porosity throughout the entire piece: This defect is caused by a lack of pressure during the rolling compaction process. In this case, the piece is discarded as it would not pass quality control.

27.4 LACK OF IMPREGNATION ON THE FIBRES

Lack of impregnation refers to areas within a composite where the fibres are not fully wetted by the resin, leading to poor load transfer between the fibre and matrix, thereby reducing overall strength. This lack of impregnation could lead to voids, as well as delamination, which can promote separation between plies under stress, and mainly compromises the composite's structural integrity and creates conditions that accelerate "resin degradation". These defects significantly reduce the fatigue life and durability of composite materials.

There are other factors that could lead to this type of defect, for example, when the resin is too viscous or has too short a gel time; another very common reason is when trying to laminate several layers at the same time in areas where access is too complicated.

Proper control during manufacturing processes like resin transfer moulding (RTM) or autoclave curing is critical to minimise lack of impregnation.

27.4.1 Resin Degradation

Resin degradation can be defined as the breakdown of the polymer matrix due to environmental, mechanical, or chemical factors, such as cyclic loading, which creates micro-cracking in the matrix and can propagate over time; the intrusion of moisture, which creates a hydrolysis effect and weakens the resin; and thermal degradation, which occurs over a long period during which the laminate has been exposed to a hostile environment and high temperatures. All these defects have several consequences, where the mechanical properties reduce their effectiveness, and the ability to protect the reinforcements is also affected and reduced, as well as delamination and crack propagation becoming more likely.

There are some measures that can be taken to help avoid resin degradation:

- Check room temperature, since low temperature during lamination could affect it directly.
- Avoid excessive loss of styrene.
- Use the proper resin, considering the external factors to which it will be exposed.

27.5 FATIGUE

Fatigue is the condition in which a material fails or cracks as a result of repeated (cyclic) stress. Therefore, fatigue is defined as the permanent, localized, and progressive structural change that takes place in a material exposed to repeated deformations. These stresses can be produced by repetitive external loads or by thermal stresses resulting from alternate heating and cooling.

The fatigue life of a material is defined as the total number of stress cycles required to cause a failure. The most common method of studying fatigue life is to use cyclic loads of constant amplitude and record the number of cycles required to reach that failure.

Therefore, the fatigue process in composites can be described by the initiation of primary cracks in the matrix, followed by fibre fracture, loss of adhesion, formation of secondary cracks, and then possible delamination.

Figure 27.12 provides a clear representation of fatigue behaviour for different materials, where maximum stress vs. number of cycles are used. Stress is normally expressed as a percentage of the material's ultimate strength (UTS), while the number of cycles before failure is expressed on a logarithmic scale to span the wide range of cycles. For example aluminium has a more aggressive decline than composites, especially at high cycles; the curve will never flatten. Kevlar has a flatter curve compared to aluminium and steel, and S-glass has a similar behaviour to Kevlar, but slightly lower fatigue resistance at equivalent stress levels.

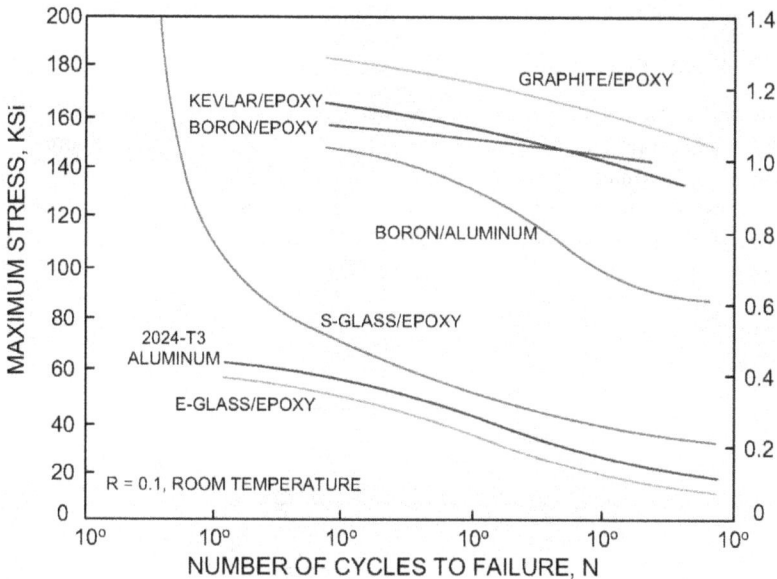

FIGURE 27.12   Representation of fatigue behaviour for different materials.

27.6   TESTING AND QUALITY CONTROL OF COMPOSITE MATERIALS

Testing and quality control are essential to ensure the performance, reliability, and durability of composite materials. These processes verify that the material meets design and application requirements and is free of defects by evaluating the mechanical, physical, and chemical properties, which will ensure a long service life.

Quality control will be carried out through standardized tests, and in some cases non-standardized tests if required. Once the tests have been performed, they will show whether the composite material part meets the quality requirements.

Three ways to test or control quality in composite materials can be identified:

- Composite test specimen.
- Destructive tests.
- Non-destructive tests (NDTs)

27.7 COMPOSITE INSPECTION METHODS

Inspection methods play a crucial role in detecting defects that could affect fatigue life. Some of the most common techniques include

27.7.1 VISUAL INSPECTION

A first and basic, but useful inspection method for composite structures is visual inspection. The operator looks for visible signs of damage to the structure, such as debonding, delamination, or any unusual aspects on the surface.

The most important advantage of visual inspection is the time. It is a very quick process and does not need special equipment, but this method fully depends on the operator's experience and skill in detection.

List of Visual Inspection Details
Bubbles
Wrinkles
Loose fibres
Resin-starved areas (lack of resin)
Resin-rich areas (excess of resin)
Ply orientation
Porosity
Surface defect
Fibre distortion
Voids
Contamination
Vacuum leak
Delamination
Ply gaps
Bridging
Debonds
Damaged plies
Micro-cracking
Uncured resin
Where they could be found?
Lay-up sequence
Bagging
Fibres
Resin
Curing process
Tool mould

27.7.2 TAP TESTING

This method is also simple; it is based on tapping a piece of metal along the surface. Again, it will be under the operator's criteria, but basically a metallic sound will indicate a good end, while a "snap" sound will indicate a defect such as delamination or debonding. As the structure becomes thicker, it becomes harder to identify, and this method would not work through the core.

27.7.3 ULTRASONIC INSPECTION

Ultrasonic inspection is another type of non-destructive testing. The ultrasonic test is based on the use of a transmitter and receiver circuit, as shown in Figure 27.13. This tool can provide the location of a possible crack, its size, and orientation. There are three types of ultrasonic inspection: A-Scan, C-Scan, and ANDSCAN.

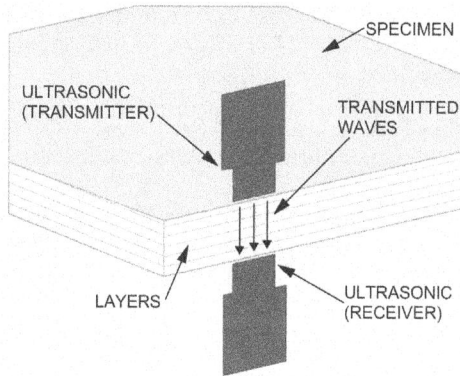

FIGURE 27.13   Representation of ultrasonic inspection method.

27.7.4 THERMOGRAPHY

Thermography is mostly used on composite structures. Using the application of heat, the part is inspected through infrared camera imaging to observe heat distribution.

27.7.5 RADIOGRAPHIC TESTING

Radiographic testing (RT) is the most commonly used testing method. For example, delamination, which is the most common type of failure in composites, can only be seen with RT.

27.7.6 X-RAY INSPECTION

X-ray inspections on composite parts are quite similar to those on metal structures, where the material density has a huge influence on the obtained result.

27.7.7 ACOUSTIC EMISSION

Acoustic emission (AE) is really effective; it is a mechanical vibration generated by the material defect, such as micro-cracking in a matrix, localized delamination, and fibre-matrix debonding.

27.7.8 SHEAROGRAPHY

It is a laser optical method in which the need for skilled users is not necessary. It is worth mentioning that this method is mainly used to detect delamination defects, as it does not have good efficiency in detecting other types of defects.

BIBLIOGRAPHY

'Analysis and Performance of Fiber Composites' (Third edition). Bhagwan D. Agarwal, Lawrence J. Broutman and K. Chandrashekhara. Wiley India Pvt. Ltd., 2015.

'Composite Materials'. S. C. Sharma. Alpha Science International, Ltd, 2000.

'Engineering Materials 1 – An Introduction to their Properties and Applications' (Second edition). Michael F. Ashby and David R. H. Jones. UK: Department of Engineering, University of Cambridge, 1996.

'Fiber Composite Analysis and Design: Composite Materials and Laminates. Volume I'. Z. Hashin, B. W. Rosen, E. A. Humphreys, C. Newton and S. Chaterjee. Washington, DC: U.S. Department of Transportation: Federal Aviation Administration Office of Aviation Research, 1997.

'Handbook: An Engineering Compendium on the Manufacture and Repair of Fiber Reinforced Composites'. R. L. Ramkumar, N. M. Bhatia, J. D. Labor and J. S. Wilkes. NJ, USA: Department of Transportation FAA Technical Center: Atlantic City International Airport, 1987.

'Handbook of Composites' (Second edition). S. T. Peters. Mountain View, CA, USA: Process Research, 1997.

'Introduction to Composite Materials Design'. Ever J. Barbero. USA: Department of Mechanical & Aerospace Engineering – West Virginia University/Taylor & Francis, 1998.

'Marine Composites' (Second edition). Eric Green Associates. MD: Eric Green Associates Inc., 1999.

'Mechanics of Composite Materials' (Second edition). Robert M. Jones. Taylor & Francis, 1999.

'Mechanics of Laminated Composite Plates and Shells - Theory and Analysis' (Second edition). J. N. Reddy. CRC Press, 2004.

'Shigley's Mechanical Engineering Design' (Tenth edition). Richard G. Budynas and J. Keith Nisbett. McGraw Hill Education, 2015.

'Structural Composite Materials'. F. C. Campbell. ASM International, 2010.

'Theory of Composites Design'. Stephen W. Tsai. Department of Aeronautics and Astronautics: Stanford University, 1992.

28 Laminate Sandwich Panels

The laminate sandwich structure panels are known as an advanced material solution primarily focused on combining lightweight constructions with high mechanical performance.

The first attempts back in the day included simple layered constructions, such as wooden shields or laminated boat hulls. However, the introduction of honeycomb as known today is a derivative procedure taken from aviation, where aluminium honeycomb cores were developed during World War II with the intention of improving stiffness on aircraft components while minimizing weight. But these early sandwich panels were mostly metal-based.

By the time that polymers and fibre reinforcement structures (carbon fibre, fibreglass, and aramid fibres) made their appearance, advances in engineering solutions incorporated these materials into face sheets, making an impressive improvement in strength-to-weight ratios.

In the modern era, sandwich panels became essential to high-performance engineering applications such as aerospace, maritime, automotive, and even renewable energy, using them in wind turbine blades.

28.1 LAMINATE SANDWICH STRUCTURE

A laminate sandwich structure consists of three main components, working together to maximise stiffness and strength. Following the exploded assembly represented in Fig. 28.1, this type of construction is performed by thin, stiff, and strong fibre

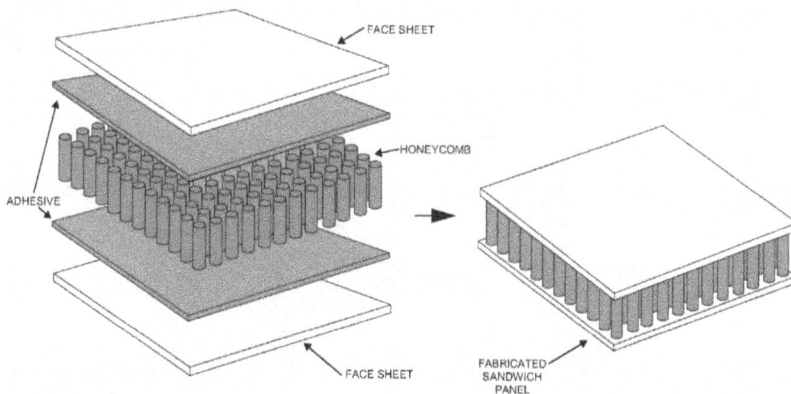

FIGURE 28.1 Representation of honeycomb laminate sandwich structure.

DOI: 10.1201/9781003565222-28

243

composite sheets separated by a thick layer of a low-density material. This thick layer of low-density material is commonly known as core material, which could be light foam, like Nomex core or Rohacell, a honeycomb, or even a corrugated core. This core material needs a joining interface that is generally adhesively bonded to the face sheets to generate the bonding between panels.

Saying it in simple words, laminate sandwich construction is a process in which a "panel sandwich" is made by a lightweight, low-density core material bonded between two skins of a laminate such as carbon fibre, fibreglass, or aramid (Kevlar).

1) **Face Sheets:** Commonly named the "outer layers", they are in charge of resisting in-plane and bending stresses. Using high-modulus materials like carbon fibre will provide stiffness to the structure, especially when the panel is subjected to bending stresses, since they work in tension and compression; while glass fibre will only offer cost-effective strength.

2) **Core:** The main function of core materials is to separate the face sheets, increasing the moment of inertia of the panel; this is a key factor when referring to stiffness, and in this way, it will absorb shear forces as well as prevent the face sheets from buckling under external load. Core materials typically used include honeycomb, foam, or corrugated materials.

3) **Adhesive or Bonding Interface:** Ensures a strong connection between the core and face sheets, maintaining structural integrity under loading conditions.

The faces of a sandwich panel handle bending moments by carrying tensile and compressive stresses, while the core manages transverse forces by resisting shear stresses. This combination of materials and layout allows the panel to behave like an I-beam, providing exceptional stiffness at a fraction of the weight. To explain this, Figures 28.2 and 28.3 will represent this comparison, where the face sheets act as the flanges of the I-beam, supporting the bending stresses (one face will be in tension, and the other in compression). Similarly, the core serves the same purpose as the web of the I-beam, transmitting the shear loads. The adhesive binds all the components together, enabling the entire structure to function as a unified system with high resistance to torsion and bending.

I - BEAM & SANDWICH PANEL

FIGURE 28.2   I-beam and sandwich panel comparison.

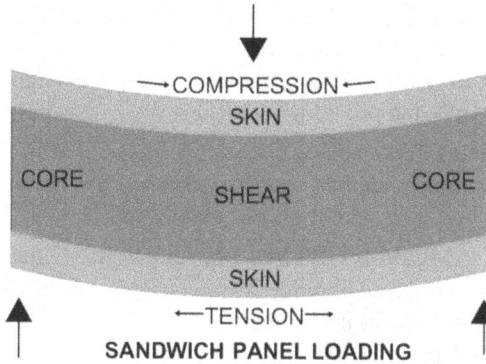

FIGURE 28.3 Representation of a loaded laminate sandwich structure.

However, establishing the strength of a composite material is a task still pending; despite the isotropic materials, the strength depends directly on the directional properties of an individual ply, which can vary along the longitudinal axis, transversely, and in shear. In addition to this, it will differ between tension and compression.

28.2 LAMINATE SANDWICH MANUFACTURING TECHNIQUES AND MATERIALS USED

By combining different materials and manufacturing techniques, laminate sandwich structures can be customized for several applications, including naval, aerospace, and automotive.

As mentioned before, different materials can be used in the outer layers of the structural sandwich panel, such as carbon fibre, which will provide high strength, stiffness, and fatigue resistance. While glass fibres could provide good corrosion resistance and would be more beneficial in terms of cost. But if great toughness and impact resistance are needed, then aramid fibres are the choice.

In terms of the core material, it is important to gain lightweight and excellent compressive and shear strength; this can be provided by different honeycomb materials such as aluminium, Nomex, or even polymers. Foam cores such as PVC, Polyurethane, or Polystyrene are excellent options for marine applications since they are easy to shape, provide insulation, and are cost-effective. At last, although not often used, balsa wood could work as a core material for some marine applications and wind turbines.

There will be no laminate sandwich structure without a proper adhesive that bonds them together; this could be epoxy-based (resin) to resist high temperatures as well as high strength, or it could be polyurethane adhesives, which provide more flexibility.

After going through the material selection process, it is crucial to combine it with the proper manufacturing method:

- **Hand Lay-Up**: Layers of fibre and resin are manually applied over a mould, while core materials are inserted before curing. This is a low-cost process but demands a qualified operator to perform it.

- **Vacuum Bagging:** After laying up the laminate and core, the assembly is enclosed in a vacuum bag. Air is removed to compact the layers and ensure proper adhesion between the core and face sheets. This improves quality by removing air voids and providing consistent resin distribution.
- **Autoclave Curing:** The sandwich assembly is placed in an autoclave, where heat and pressure are applied to cure the resin. This ensures a void-free, high-strength structure ideal for high-performance applications.
- **Resin Transfer Moulding (RTM):** Dry fibre reinforcements and the core are placed into a closed mould. Resin is injected under pressure, saturating the fibres and bonding the layers. This provides a high-quality finish compared to previous processes and is efficient for high-volume production.
- **Vacuum-Assisted Resin Transfer Moulding (VARTM):** Similar to RTM, but resin infusion is assisted by vacuum pressure, ensuring thorough saturation of fibres and core. This is lower cost than traditional RTM and is mostly used on large structures.
- **3D Printing:** Introducing some new technology into composite manufacturing, core materials with intricate geometries are created using 3D printing, and then face sheets are applied afterward. The main advantage of this is the ability to form complex core designs, including material waste reduction.

28.3   ADVANTAGES AND DISADVANTAGES OF COMPOSITE LAMINATE SANDWICH STRUCTURES

Composite laminate sandwich structures offer several advantages, but they also have limitations depending on the materials, design, and applications

28.3.1   Advantages of Laminate Sandwich Structure

One of the main advantages of sandwich construction built with composite face sheets is that the core material can be bonded or co-cured together with the face sheets. In general, a sandwich construction has the following advantages:

- High ratio of bending stiffness to weight compared to monolithic laminate composites.
- High resistance to mechanical fatigue.
- The sandwich structure resists bending and provides high stiffness due to the separation of the face sheets by the core.
- Good damping characteristics.
- Improved thermal isolation.
- The materials and thicknesses can be customized to specific application requirements.

28.3.2 Disadvantages of Laminate Sandwich Structure

Composite laminate sandwich structures, despite their many advantages, also have several limitations:

- Complex Manufacturing: Advanced fabrication methods require specialised equipment and expertise.
- When moisture is trapped in the core material, it may cause corrosion problems.
- Delamination Risks: Improper bonding between face sheets and the core can lead to delamination, reducing structural integrity.
- A well-performed quality control process is needed during fabrication to ensure there is no de-bonding in the adhesive layer.
- If failure occurs due to debonding, it could start to propagate in the adhesive layer during service and affect the efficiency under loads.
- Difficulty in Repairs: Damage to the core or face sheets can be challenging and costly to repair compared to traditional materials like metals.
- Limited Recycling Options

Core Vulnerability: The lightweight core can be susceptible to damage from concentrated loads or impacts, leading to core crushing or shear failure.

28.4 FAILURE OF SANDWICH PANEL STRUCTURES

Failure modes in sandwich structures are very different from those in a monolithic laminate structure. They can occur due to various mechanisms, depending on the loading conditions, material properties, and environmental factors. Understanding these failure modes is critical when designing composite structures.

To mention the most common failure mechanisms in composite laminate sandwich structure panels:

28.4.1 General Buckling

General buckling is a failure mode that occurs when a composite laminate panel structure, subjected to compressive or shear loads, suffers deformation in its plane due to instability. Figure 28.4 shows a general buckling of the sandwich panel. This type of failure typically happens when the applied load exceeds the panel's critical

28.4   General buckling of a panel.

buckling load. It might occur due to the lack of thickness in the laminate sheets or low density (core rigidity) of the selected core material.

In some applications, composite panels are designed to tolerate post-buckling loads, especially in aerospace structures.

28.4.2 SHEAR CRIMPING

Following general buckling, shear crimping particularly occurs in sandwich structures with a lightweight core. It is often associated with thin cores or cores made from low-strength materials like foams or low-density honeycombs, typically with low shear modulus or low adhesive shear strength. It occurs when the core material is unable to support the shear loads transmitted by the face sheets, causing localized deformation or crumpling of the core as shown in Figure 28.5.

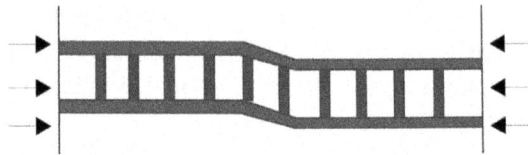

FIGURE 28.5 Sandwich panel shear crimping.

28.4.3 FACE WRINKLING

Face wrinkling is a failure mode in composite laminate sandwich panels where the face sheet deforms into a wavy pattern (buckles), often caused by compressive or bending loads. Figure 28.6 shows a representation of the sandwich panel face wrinkling. The wrinkling of the composite face sheet can be inwards or outwards depending on the core compression strength and adhesive strength in tension. This type of failure compromises the panel's structural integrity and can propagate, leading to further damage such as core crushing or delamination.

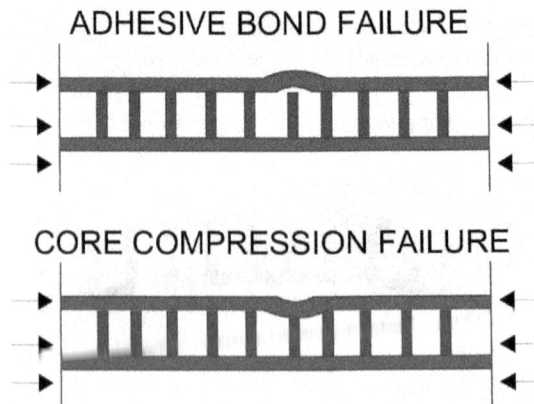

ADHESIVE BOND FAILURE

CORE COMPRESSION FAILURE

FIGURE 28.6 Sandwich panel face wrinkling.

28.4.4   INTRACELL BUCKLING

This is a localized failure mode where the face sheet undergoes buckling within individual cells of the core structure. This is well represented on Figure 28.7 where different cells experience individual buckling. It occurs in panels with cellular cores due to the presence of thin composite face sheets or large core cell sizes.

FIGURE 28.7   Sandwich panel intracell buckling.

It is particularly associated with sandwich panels that use honeycomb or other cellular core materials and typically occurs when compressive or bending loads affect the unsupported regions of the face sheet over the core cells.

28.4.5   FACE SHEET FAILURE

Face sheet failure occurs when the face sheets, which are primarily responsible for carrying bending and in-plane loads, are unable to withstand the applied stresses. Figure 28.8 represents a tensile failure in facing. This failure mode can appear in various forms, depending on the type of load, material properties, and panel design, and is typically caused by the lack of panel thickness, face sheet thickness, or face sheet strength.

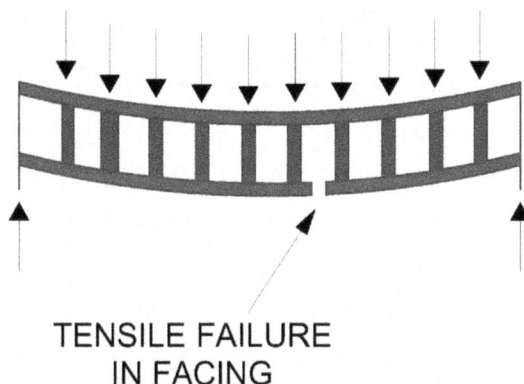

TENSILE FAILURE
IN FACING

FIGURE 28.8   Sandwich panel tensile failure in facing.

28.4.6 Transverse Shear Failure

Transverse shear failure is caused by the lack of core shear strength or panel thickness, and it happens when the core material is unable to resist shear stresses transferred from the face sheets. This failure mode is particularly significant in panels exposed to transverse loads or bending, as shown in Figure 28.9, as the core is responsible for carrying shear forces while the face sheets handle bending stresses.

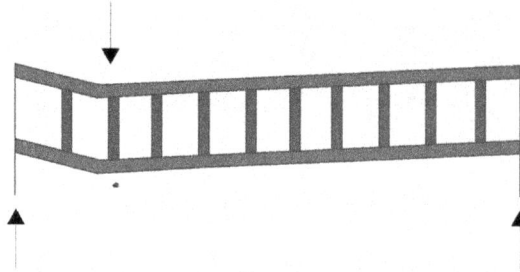

FIGURE 28.9   Sandwich panel transverse shear failure.

28.4.7 Flexural Crushing of Core

Flexural crushing of the core occurs in composite laminate sandwich panels when the compressive stresses induced by bending loads exceed the compressive strength of the core material. This is basically caused by the lack of core compressive strength or an excess of panel deflection. Figure 28.10 represents a flexural crushing of core failure within a sandwich panel.

FIGURE 28.10   Sandwich panel flexural crushing of core.

This failure mode is critical in applications where sandwich panels are exposed to high bending moments, as it compromises the structural integrity and the capacity of the panel to carry loads.

28.4.8 LOCAL CRUSHING OF CORE

This failure is caused by low core compressive strength. Local crushing of the core occurs when concentrated loads or impacts cause the core material to deform or fail at a specific point. Figure 28.11 shows a centered core deformation due to a local crushing of core within a sandwich panel. Unlike flexural crushing, which arises from bending-induced compressive stresses, local crushing is driven by localized forces that exceed the core's compressive strength, often near load application points or under fixtures.

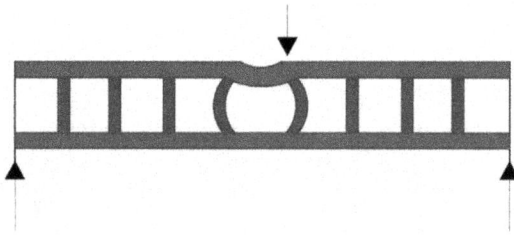

FIGURE 28.11 Sandwich panel local crushing of core.

28.5 TYPICAL PROPERTIES OF SANDWICH CORE MATERIALS

As previously explained, sandwich core materials provide structural support, improve bending stiffness, and reduce overall weight. Since the core materials sit between the face sheets and are responsible for carrying shear loads, dampening vibrations, and maintaining the shape of the panel, the properties of core materials are strictly related to the performance of sandwich panels in terms of strength, stiffness, weight, impact resistance, and durability. Below are the typical properties of common sandwich core materials:

Core Material	Specific Gravity	Shear Modulus		Shear Strength		Through-Thickness Young's Modulus		Through-Thickness Compressive Strength	
		Absolute value	Specific value	Absolute value	Specific value	Absolute value	Specific value	Absolute value	Specific value
Pvc foam	0.075	25	320	0.8	10.7	50	667	1.1	15
Pvc foam	0.130	40	308	1.9	14.6	115	885	3.0	23
Pvc foam	0.190	50	260	2.4	12.6	160	842	4.0	21
Pu foam	0.100	10	100	0.6	6.0	39	390	1.0	10
Pu foam	0.190	30	158	1.4	7.4	83	437	3.0	16
Synthetic foam	0.400	430	1070	-	-	1200	3000	10	25
Synthetic foam	0.800	1000	1250	21	26	2600	3250	45	56
End-grain balsa	0.100	110	1100	1.4	14	800	8000	6	60
End-grain balsa	0.180	300	1670	2.5	14	1400	7780	13	72
Aluminium honeycomb	0.070	455/205	6500/2930	2.2/1.4	31/20	965	13790	3.5	50
Aluminium honeycomb	0.130	895/365	6885/2810	4.8/3.0	37/23	2340	18000	9.8	75
Grp honeycomb	0.080	117/52	1462/650	2.3/1.4	29/18	580	7250	5.7	71
Aramid paper	0.065	53/32	815/492	1.7/1.0	26/15	193	2970	3.9	60

The relationship between mechanical properties such as shear strength, compressive strength, and density for core materials can also be reflected in a graph, similar to the ones presented in Figure 28.12 and Figure 28.13. These graphs are based especially on shear strength or compressive strength, always related to the material density.

CORE MATERIALS

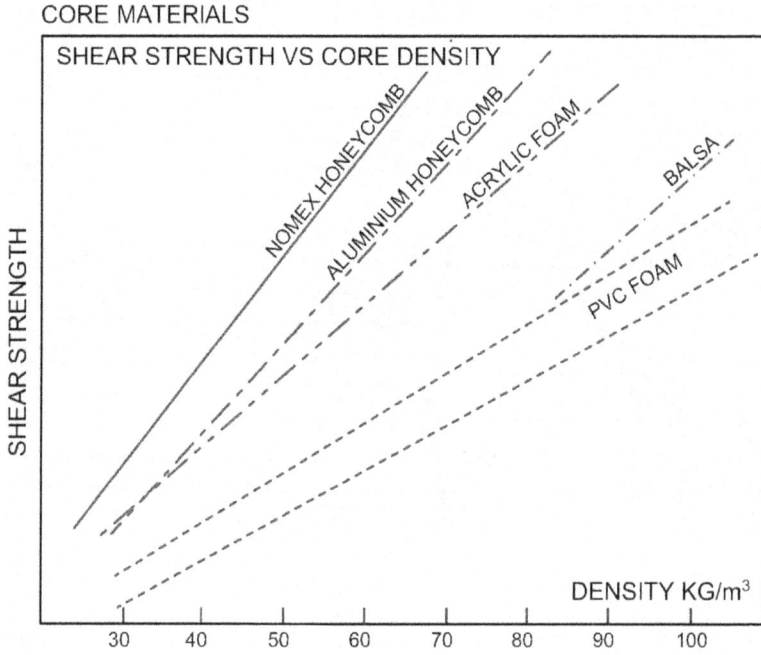

FIGURE 28.12   Shear strength vs core density.

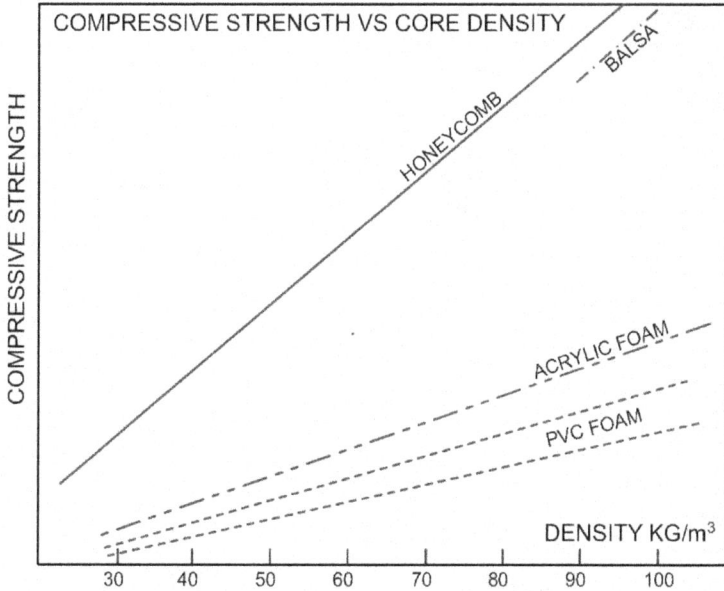

FIGURE 28.13   Compressive strength vs core density.

As density increases, shear strength generally increases because denser materials have a more robust internal structure to resist deformation. For honeycomb cores, for example, shear strength is influenced by cell size, wall thickness, and the material itself. However, higher density in honeycomb structures shows higher shear strength.

On the other hand, PVC foams show a non-linear increase in behaviour; this is due to changes in the foam's microstructure as density rises, reducing voids and increasing load-bearing material.

Before analysing the compressive strength vs. density graph, it is important to remind that compressive strength is the capacity of a material to withstand axial compressive forces, and as higher density typically results in higher compressive strength, since it contains more material per unit volume to resist the loads.

The compressive strength in honeycombs will be managed by the ability of the cell structure to withstand and resist buckling. So, high-density honeycombs are less likely to suffer local failures. On balsa wood, for example, the compressive loads are transferred by its uniform and grain-aligned structure, which has a direct relationship between compressive strength and density. PVC foams behave in a similar way, but their compressive strength will purely depend on how the polymer structure adapts to the increased density.

BIBLIOGRAPHY

'Analysis and Performance of Fiber Composites' (Third edition). Bhagwan D. Agarwal, Lawrence J. Broutman and K. Chandrashekhara. Wiley India Pvt. Ltd., 2015.

'Handbook: An Engineering Compendium on the Manufacture and Repair of Fiber Reinforced Composites'. R. L. Ramkumar, N. M. Bhatia, J. D. Labor and J. S. Wilkes. NJ, USA: Department of Transportation FAA Technical Center: Atlantic City International Airport, 1987.

'Marine Composites' (Second edition). Eric Green Associates. MD: Eric Green Associates Inc., 1999.

'Mechanics of Composite Materials' (Second edition). Robert M. Jones. Taylor & Francis, 1999.

'Sandwich Structural Composites – Theory and Practice'. W. Ma and R. Elkin. CRC Press, 2022.

'Structural Composite Materials'. F. C. Campbell. ASM International, 2010.

'Theory of Composites Design'. Stephen W. Tsai. Department of Aeronautics and Astronautics: Stanford University, 1992.

29 Nomenclature

In composite laminates, fibre orientations and stacking sequences are described relative to a coordinate system, typically based on the principal axes of the structure, as shown in Figure 29.1. Each ply in the laminate is characterized by its orientation angle, which describes the fibre alignment relative to the 0 degrees axis.

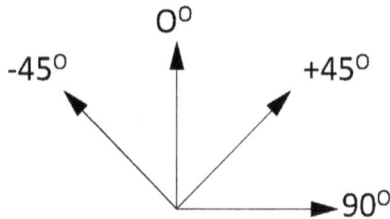

FIGURE 29.1 Representation of angle laminate orientations on a coordinate system.

There is more than one way to represent the laminate stacking sequence. After the 0 degrees fibre direction has been defined by the x-axis, the plies that are not aligned to that x-axis must be assigned an angle.

Clockwise rotations are referred to as positive angles, and counterclockwise rotations are referred to as negative angles.

Furthermore, composite lay-ups are represented using shorthand notations to describe the stacking sequence and fibre orientation of each ply. The proper and general way to describe the fibre orientations of all the plies that fully conform to a laminate or lay-up is to write them sequentially within brackets, separated by a slash.

To explain this quickly, the plies having fibres oriented at 45 degrees may have fibres in a +45 or −45 degrees direction, all referred to the main axes of the laminate. The use of the ± prefix implies that there are two plies, one having fibres along +45 and the other along −45. Figure 29.2 shows a stack of plies using ±30 and ±45, and a proper description is used to denote the stack of plies for this particular laminate. In this case, a subscript "T" is also used, which is placed after the closing bracket to denote the "Total laminate".

DOI: 10.1201/9781003565222-29

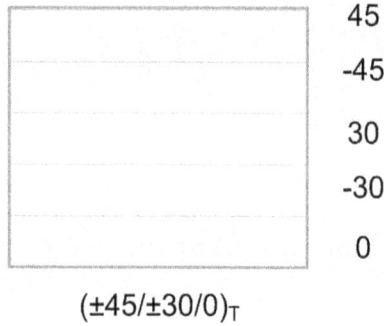

45

-45

30

-30

0

$(\pm 45/\pm 30/0)_T$

FIGURE 29.2   Sample of stack of plies using ± notations and total laminate indicator.

For laminate sequences using a subscript "S" after the closing brackets, as shown in the laminate description of Figure 29.3, it will denote the existence of symmetry, which means that it is a symmetric laminate.

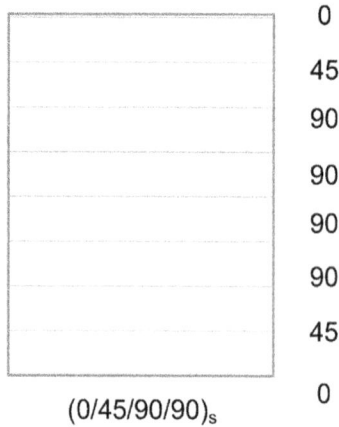

0

45

90

90

90

90

45

0

$(0/45/90/90)_s$

FIGURE 29.3   Sample of stack of plies using "S" subscript notation to denote symmetry.

29.1   STACKING OF LAMINATE SHORTHAND EXAMPLES

$\left[0° \ / \ 90° \ / +45° \ / -45°\right]_S$ indicates a symmetric lay-up, and it represents the following laminate:

$$\left[0° \ / \ 90° \ / +45° \ / -45° \ / \ -45° \ / +45° \ / \ 90° \ / \ 0°\right]$$

A similar situation, but imagine having an underline or an overbar in one of the ply indicators, $\left[0° \ / 90° \ / +45° \ / \underline{-45°}\right]_S$ (or overbar) indicates a symmetric lay-up through the last ply centre surface, as represented in Figure 29.4:

$$\left[0° \ / 90° \ / +45° \ / -45° \ / \ +45° \ / \ 90° \ / \ 0°\right]$$

−45 degrees will work as a centred ply for the mirrored symmetry.

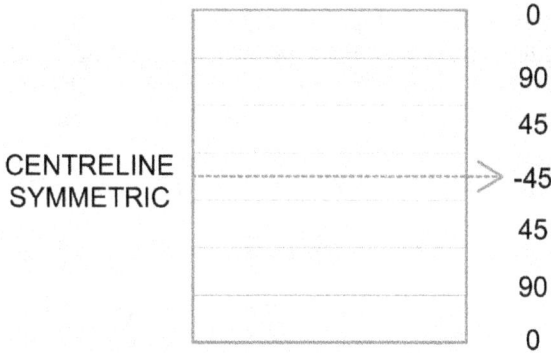

FIGURE 29.4 Sample of a centreline symmetric stack of plies.

If an "as" subscript is used, $\left[0° \ / 90° \ / +45° \ / -45°\right]_{as}$ indicates that is an asymmetric lay-up:

$$\left[0° \ / 90° \ / +45° \ / -45° \ / \ +45° \ / \ -45° \ / \ 0° \ / \ 90°\right]$$

It means that each ply will have its own symmetric partner, but not necessarily in a mirrored order as in the previous sample.

Using a number as a subscript, in the following sample expressed with the number six,

$\left[+45°/ -45°\right]_6$ indicates a laminate of +45/−45 repeated six times stacked one on top of the other:

$$\left[+45°/ -45°/+45°/ -45°/+45°/ -45°/+45°/ -45°/+45°/ -45°/+45°/ -45°\right]$$

All these different notations are really useful for expressing laminate sequences without writing long chains of plies. A combination of different definitions can be used, for example:

$$\left[\left[0° \ / 90°\right]_4 \left[55° \ / -55°\right]_4\right]_S$$

A fabric or woven material could be shown sometimes as $(30° \,/\, 60°)$ or $(0° \,/\, 90°)$, or $(+45° \,/ -45°)$, or (±45)

When mixed materials are used, they may be defined with the letters (a) and (b): And it would be shown like this:

$$\left[+45°_a \,/ -45°_a \,/\, 0°_b \right]_s$$

29.2 BALANCED LAY-UP

To identify a balanced lay-up, it is important to understand the following rule: it is referred to a lay-up that has been made by the corresponding +angle/-angle or 0/90 plies. A balanced lay-up ensures that off-axis layers are paired with equal and opposite orientations referred to each ply, but not necessarily or with symmetry.

$$\left[0° \,/\, 90° \,/ -35° \,/ +35° \,/ +15° \,/ -15° \,/\, 90° \,/\, 0° \right]$$

or

$$\left[90° \,/\, 0° \,/\, 90° \,/ -35° \,/ +35° \,/ -15° \,/\, 0° \,/ +15° \right]$$

Note that the relative position of each pair does not matter; what really matters is that each ply has its equal and opposite. For example, for each ninety degrees, there is a zero degrees; for a minus 35 degrees, there is a positive 35 degrees, and the same with the fifteen degrees.

Balancing lay-ups reduces coupling effects such as twisting or bending under in-plane loads. This is crucial for laminates subjected to multidirectional stresses.

29.3 SYMMETRIC LAY-UP

A symmetric lay-up ensures that the laminate sequence is mirrored about its mid-plane; it will have strict symmetry of plies about the mid-surface.

Symmetry prevents out-of-plane deformations, like warping, due to in-plane loads or thermal effects.

$$\left[0° \,/\, 90° \,/ +45° \,/ -45° \,/ -45° \,/ +45° \,/\, 90° \,/\, 0° \right]$$

In the case presented above, the lay-up is symmetric and balanced. However, a lay-up can be symmetric but unbalanced.

$$\left[0° \,/\, 0° \,/ +45° \,/ -45° \,/ -45° \,/ +45° \,/\, 0° \,/\, 0° \right]$$

A laminate is symmetric (mirrored about the mid-plane) but unbalanced when it does not have equal $+\theta°$ and $-\theta°$ plies. This unbalance in the laminate can cause coupling effects when exposed to axial loads.

If we think specifically about a laminated plate, unbalanced and asymmetrical laminates will tend to bend when exposed to load, as well as they could also lead to twisting.

BIBLIOGRAPHY

'Analysis and Performance of Fiber Composites' (Third edition). Bhagwan D. Agarwal, Lawrence J. Broutman and K. Chandrashekhara. Wiley India Pvt. Ltd., 2015.

'Analysis of Composite Materials. A Survey'. Z. Hashin. *Journal of Applied Mechanics*, Vol. 50 (1983), No. 3, 481–505.

'Basic Mechanics of Laminated Composite Plates'. A. T. Nettles. Alabama: Marshall Space Flight Center-MSFC (NASA), 1994.

'Composite Materials'. S. C. Sharma. Alpha Science International, Ltd, 2000.

'Handbook: An Engineering Compendium on the Manufacture and Repair of Fiber Reinforced Composites'. R. L. Ramkumar, N. M. Bhatia, J. D. Labor and J. S. Wilkes. NJ, USA: Department of Transportation FAA Technical Center: Atlantic City International Airport, 1987.

'Handbook of Composites' (Second edition). S. T. Peters. Mountain View, CA, USA: Process Research, 1997.

'Introduction to Composite Materials Design'. Ever J. Barbero. USA: Department of Mechanical & Aerospace Engineering – West Virginia University/Taylor & Francis, 1998.

30 Composite Recycling

Recycling composite materials has become a critical matter over the last decade due to the increasing use of composites in several industries, as well as the need to address their environmental impacts.

In the mid-20th century, the first development of composite materials such as fibre-reinforced polymers was mainly thought for high-performance applications, particularly in the aerospace and defence industries, where lightweight, strength, and durability are essential. This achievement was accomplished by a unique combination of a reinforcement material (glass or carbon fibres) with a matrix material (epoxy or polyester resins).

By the late 20th century, the need for sustainable end-of-life (EOL) management of composites became a key factor to solve when looking to the future. The disposal of composite waste in landfills became a major problem that created environmental hazards and resulted in the loss of the economic potential of the embedded fibres. Furthermore, the increasing pressure to apply proper regulations, together with public awareness of environmental issues, helped to emphasize the importance of recycling.

The growing use of composites and the requirements of end-of-life (EOL) regulations call for a deeper focus on recycling and investment in global supply chains. Despite this need, the advancement in terms of commercializing recycling technologies and expanding recycling capacity has been quite slow to date.

30.1 RECYCLING SUSTAINABILITY

Various efforts are underway to make recycling feasible. Some companies are innovating to develop more efficient techniques, while others focus on processing composite waste into new products and increasing their operational capacity. Meanwhile, an increasing number of projects are highlighting the production of recyclable goods, often incorporating closed-loop recycling systems.

Recycling, alongside biomaterials and energy-efficient solutions, plays a key role in sustainability but also involves some challenges. To start, composites have a complex material structure; they combine fibres and resin in a tightly bonded structure, making it hard to achieve any potential separation and recovery system. This also involves material properties, where the structural and mechanical integrity of recycled fibres is difficult to maintain, not to mention the cost of recycling composites, which often incurs higher costs than producing the actual materials. At last, it is hard to even think of a standardised method for recycling composites; the variety of materials complicates the development of universally applicable recycling methods.

DOI: 10.1201/9781003565222-30

30.2 RECYCLING METHODS APPLIED FOR COMPOSITE MATERIALS

The first methods used to dispose of composite waste were incineration and land-filling. Both are widely categorized as harmful to the environment; incineration releases greenhouse gases and other pollutants, while landfilling poses similar risks and offers only a temporary solution for materials designed to last for decades. Other techniques combine combustion and incineration to convert waste into heat for energy recovery, such as electricity production. However, these methods still produce pollutants, including ash as a byproduct.

Several techniques have been developed and refined since then to address the challenges of composite recycling. They can be broadly classified into Mechanical, Thermal, and Chemical processes, with the main goals being

- Material Recovery: Recover the fibres and resin, where possible.
- Waste Reduction: Reduce the volume of composite materials sent to landfills.
- Energy Efficiency: Minimise the energy used in recycling processes.
- Economical: Develop methods that can be competitive with newly produced materials in terms of material cost production.

30.2.1 MECHANICAL RECYCLING

In mechanical recycling, composite materials are physically broken down into smaller particles. The process involves a number of steps, which typically begin with shredding the composite into small pieces, reducing this waste into manageable pieces. These pieces are then ground into fine powders or particles using specialised milling equipment. In the end, the resulting material, which is important to remind does not retain the original properties, can serve as fillers or reinforcements in new composite formulations.

Mechanical recycling is relatively simple and cost-effective compared to other methods; it has minimal environmental impact since it does not require high energy inputs or chemical treatments. However, it often leads to a reduction in the structural integrity of the recovered material, a process known as downcycling, limiting and reducing the potential reuse to lower-value applications.

30.2.2 THERMAL RECYCLING

Thermal recycling methods utilise heat to decompose composite materials, primarily targeting the recovery of fibres while sacrificing the resin matrix. Two prominent techniques within this category are pyrolysis and combustion.

30.2.2.1 Pyrolysis

This technique is normally performed in a high-temperature chamber and involves heating the composite material in an oxygen-free environment. This controlled heating causes the resin matrix to break down into gaseous and liquid byproducts, leaving behind the reinforcement fibres. Pyrolysis is particularly effective for recovering high-quality carbon fibres, which retain most of their mechanical properties. However, the process is energy-intensive, and the byproducts require careful management to mitigate environmental impacts.

30.2.2.2 Combustion

Combustion, on the other hand, involves burning composite materials in the presence of oxygen; its primary goal is energy recovery rather than material preservation.

While this method can be effective in terms of energy output, it destroys the reinforcement fibres and generates significant emissions, leading to the necessity of more severe controls to minimise environmental harm.

30.2.3 CHEMICAL RECYCLING

Chemical recycling techniques encourage recovery of both the fibres and the resin matrix by using chemical reactions or solvents. This approach is the closest to producing high-quality materials suitable for reuse in advanced composite applications.

There are two types of techniques that can be performed in this category: solvolysis and depolymerization.

30.2.3.1 Solvolysis

The solvolysis process is based on dissolving the resin matrix; to do this, a combination of solvents needs to be utilised. The process consists of immersing the composite waste in a solvent mixture under specific temperature and pressure conditions, causing the resin to break down and separate from the fibres. This method excels in retaining the quality of recovered fibres and, in some cases, enables partial recovery of resin for reuse. However, this process is expensive and requires meticulous management of the chemicals involved.

30.2.3.2 Depolymerization

Depolymerization focuses on breaking the resin matrix into its original monomers using heat, catalysts, or other chemical agents. After that, these monomers can be reused to produce new resins, while the fibres remain intact. Although this method offers the potential for complete material recovery, it is a complex procedure and the costs are very high. Not to mention that it requires advanced infrastructure and expertise.

30.2.4 EMERGING METHODS

Innovative techniques and procedures are constantly under investigation and being developed to improve the efficiency and effectiveness of composite recycling. Among these, microwave pyrolysis is gaining attention for its ability to provide uniform heating as well as reduce energy consumption.

Another promising method involves supercritical fluids, which use high-pressure and high-temperature conditions to break down the resin matrix in the most efficient way while preserving the fibres.

Additionally, enzymatic recycling represents a novel frontier, where specialised enzymes selectively degrade the resin, leaving the fibres intact.

All these different emerging methods are still in the development stages but hold significant promise for the future of composite recycling.

30.3   FUTURE OF RECYCLING COMPOSITE MATERIALS

The growing use of composite materials and the importance of introducing stricter end-of-life (EOL) regulations highlight the importance of improving recycling processes and investing in global supply chains. While efforts to commercialize recycling technologies and expand recycling infrastructure have faced challenges, recent developments indicate an important acceleration in progress, but there is still more to do.

The future of composite recycling will depend on several factors, such as current technological advances, which will focus on improving methods for fibre and resin recovery. Regulatory support is also essential; policies need to promote recycling and penalize landfilling, and economic incentives could really help in this push to encourage industries to be more conscious about investing in recycling processes for composite materials.

Composite recycling is a critical aspect of modern material management, offering solutions to environmental and economic challenges. Continued innovation and collaboration will be key to advancing this field. Unity, collaboration, and partnerships between companies and industries are potential allies in this fight to avoid depending entirely on government regulations.

BIBLIOGRAPHY

'Analysis and Performance of Fiber Composites' (Third edition). Bhagwan D. Agarwal, Lawrence J. Broutman and K. Chandrashekhara. Wiley India Pvt. Ltd., 2015.
'Handbook of Composites' (Second edition). S. T. Peters. Mountain View, CA, USA: Process Research, 1997.

31 New Technologies Applied to Composite Materials

Nobody can deny that composite materials have significantly transformed industries over the last decades, from aerospace and automotive to marine, construction, and healthcare.

As shown in previous chapters, these massive improvements in applying composites to different industries were due to several factors, such as their superior strength-to-weight ratio, corrosion resistance, and design flexibility. However, modern technology has brought huge advances, and they are further revolutionizing how composites are manufactured, processed, and utilised. Automation, new material formulations, 3D printing, and more are some of the cutting-edge technologies shaping the field and creating new paths.

31.1 AUTOMATION IN COMPOSITE MANUFACTURING

Automation in composite manufacturing involves using advanced technologies and systems to simplify and improve the process of producing composite materials as well as their components, reducing manual labour, increasing production efficiency, and ensuring higher consistency and quality in the final products.

The adoption of automation in composite manufacturing has replaced many traditional manual processes, offering a number of benefits. Robotic lay-up, automated CNC machining, automated inspection controls, and automated curing systems are some of the applications that are already settled and will definitely stay in the composite manufacturing industry.

31.1.1 Robotic Lay-up Systems

A robotic lay-up system automates the process of laying composite material layers onto a mould or tool to create the desired part. These systems typically use articulated robotic arms equipped with specialized end heads or tools that help to handle composite materials, such as pre-impregnated (prepreg) fibres, dry fibres, or resin-infused fabrics. These systems are designed to follow precise paths and angles, ensuring that the fibre orientations meet the specified design requirements.

Some robotic lay-up systems also implement auxiliary technologies, such as automated fibre placement (AFP) and automated tape laying (ATL), which are widely used for fabricating large-scale composite structures. These robotic systems enable precise placement of composite materials, adding layers with high precision by using

DOI: 10.1201/9781003565222-31

narrow strips of tape or fibre, reducing material waste and ensuring uniformity. For example, AFP has been integrated into the production of aerospace components like fuselages and wings.

31.1.2 AUTOMATED CNC MACHINING, CUTTING, AND PREFORMING

Automated cutting and preforming are transforming composite manufacturing by combining precision and efficiency, ensuring precise shaping and alignment of composite materials before their final assembly or curing. As these systems continue to evolve, they will play a critical role in meeting the growing demand for lightweight, high-performance composite materials.

Automated cutting refers to the use of robotic or CNC-based systems to cut composite materials into precise shapes and sizes. Advanced cutting techniques, like laser cutting, waterjet cutting, or knife cutting, ensure the material's structural integrity and dimensional accuracy.

Preforming involves arranging and assembling the cut composite materials into a predefined shape or configuration. This step is essential when preparing materials for moulding or curing processes, such as resin transfer moulding (RTM) or autoclave.

31.1.2.1 Advantages of Automated Cutting and Preforming

- High Precision and Accuracy: Ensures exact cuts and preform shapes, meeting smaller tolerances and reducing the risk of human error, especially for complex geometries.
- Increased Efficiency: Reduces cycle times and accelerates production rates.
- Material Optimisation: Advanced nesting algorithms minimise scrap and maximise material usage.
- Improved Quality: Consistency in cutting and preforming ensures better mechanical properties and part reliability.
- Cost Savings: Reduced material waste and labour costs over time, which lead to a faster production cycle.

31.1.3 AUTOMATED INSPECTION AND QUALITY CONTROL

Automated inspection involves the use of machines and software to evaluate composite materials and products for defects or deviations from specifications, such as voids, misalignments, or delamination. It replaces manual inspections with more precise and efficient processes through the use of advanced technologies like robotics, sensors, artificial intelligence (AI), and non-destructive testing (NDT) methods to detect defects, measure dimensions, and validate the quality of parts during and after production.

As explained and shown in previous chapters, quality control (QC) involves a wide range of stages and techniques, including inspection and monitoring of manufacturing processes. Automated QC systems ensure consistent quality across large production volumes by integrating real-time data collection and analysis.

31.1.4   Automated Curing Systems for Composite

Automated curing systems use modern technologies to control and optimize the curing process of composite materials, which are essential for producing high-quality composite materials while meeting the demands of modern manufacturing.

Managing key parameters such as temperature, pressure, and curing time, and integrating them into the workflow will reduce the need for manual intervention as well as ensure consistent quality across components, together with efficiency and sustainability, making them indispensable in the production of high-performance composite components.

31.1.4.1   Components of Automated Curing Systems

Automated curing systems in composite manufacturing are composed of several components that work together to ensure precise and efficient curing of composite materials.

The main components are the heating elements, which provide the necessary thermal energy to initiate and complete the curing process. These heating components can take the form of autoclaves, industrial ovens, or induction heaters, depending on the specific application and material requirements.

Pressure control systems are another vital component, particularly in processes requiring the removal of air and ensuring even resin distribution within the composite. These systems include vacuum pumps and pressure regulators that create a controlled environment during the curing cycle.

Sensors and monitoring devices are essential too; they provide real-time feedback, enabling precise adjustments to maintain optimal curing conditions. To do this, thermocouples and other temperature sensors are strategically placed to monitor heat distribution, while some systems also incorporate humidity sensors to control the moisture content, which can affect the curing process.

The control software forms the brain of the automated curing system. This software is responsible for managing and coordinating the various parameters of the curing process, such as temperature ramp rates, hold times, and pressure values. It allows operators to program curing cycles according to the specific requirements of the composite material and part being produced.

31.1.5   Benefits of Using Automation in Composite Manufacturing

By adopting automation, manufacturers within industries can stay competitive, meet high-quality standards, and achieve sustainable growth.

- **Increased Production Efficiency:** Automation can accelerate manufacturing processes, reducing the time-consuming tasks such as trimming, laying-up, mould manufacturing, and curing processes.

  They also benefit from consistency and repeatability, working at constant speeds without fatigue through numerous repeating cycles, maintaining high productivity levels.

- **Ensures Quality and Precision:** As mentioned, repeatability and automated systems ensure uniformity and accuracy across all produced components, especially if intricate patterns and shapes are needed. This is the perfect way to reduce major defects, which sometimes represent a real challenge to achieve manually.

- **Cost Savings:** Automation reduces the costs of manual labour for repetitive and time-consuming tasks, as well as for material waste, providing high precision and material optimisation.

- **Improved Safety:** Automation minimises the exposure of workers to hazardous materials, high temperatures, and dangerous tools.

- **Sustainability:** Reduced waste by reducing material wastage, leading to lower environmental impact, as well as reduced overall consumption, optimizing the energy used.

- **Customisation and Flexibility:** Automated systems can be programmed and aligned to any specific requirements from clients, offering flexibility in production without compromising efficiency.

31.2 INTEGRATION OF NEW MATERIALS

New formulations and advancements in terms of materials have expanded the performance capabilities and applications of composite materials. This includes not just new materials in terms of fabrics but also new resin systems, hybrid composites, manufacturing processes, sustainability, and recycling.

31.2.1 HIGH-PERFORMANCE FIBRES

Next-generation fibres such as carbon nanotubes, graphene, and aramid fibres are being incorporated into composites with the goal of improving mechanical and thermal properties. Carbon nanotubes and graphene offer superior strength-to-weight ratios, electrical conductivity, and thermal stability, while reinforced polymers are in the search for incorporating nanoparticles or advanced fillers to improve toughness and reduce brittleness. These materials will provide exceptional strength, conductivity, and heat resistance.

31.2.2 SUSTAINABLE AND BIO-BASED RESINS AND FIBRES

The development of bio-based resins and natural fibres like flax and hemp has moved forward, aligning with the growing environmental concerns. These sustainable composites reduce carbon footprints while maintaining desirable properties for various applications.

31.2.3 MULTI-FUNCTIONAL MATERIALS

Smart composites embedded with sensors and actuators are emerging for applications requiring self-healing, damage detection, or adaptive performance. For example, piezoelectric composites can generate electrical signals in response to mechanical stress, enabling real-time monitoring of structural integrity.

31.2.4 Multi-Materials

Thinking about the definition of "hybrid", it is now possible to create a hybrid composite material by combining fibres, such as carbon and glass, to optimise strength, weight, and cost. Applying layers of different material types will improve properties in specific regions of the manufactured component.

It is also beneficial, but not fully recommended, to include metal components in composites for better functionality, such as conductivity or wear resistance. There is still more work to do in terms of proper bonding between these materials, but this will definitely improve their capabilities.

31.3 ADVANCED PROCESSING TECHNIQUES

New processing methods are being applied in many industries, changing the efficiency and quality of composite manufacturing.

31.3.1 Resin Transfer Moulding (RTM) with Automation

Advancements in RTM include automated injection systems and optimized mould designs that improve resin flow and reduce cycle times. These developments make RTM suitable for high-volume production.

31.3.2 Out-of-Autoclave Curing

Out-of-autoclave curing techniques can eliminate the need for expensive autoclaves while maintaining high-quality results. To name a few, vacuum-assisted resin transfer moulding (VARTM) and press curing are the most popular in this field.

31.3.3 Microwave and Induction Heating

Innovative heating techniques, including microwave and induction heating, offer rapid and uniform curing of composite materials. These methods significantly reduce energy consumption and processing time.

31.4 RAPID PROTOTYPING AND 3D PRINTING

Rapid prototyping (RP) and 3D printing technologies are transforming composite manufacturing by enabling complex geometries, customisation, and material efficiency.

This new era has reshaped the design and manufacturing landscape by enabling faster and more efficient production of complex components. Aerospace, automotive, and marine industries are in need of increasing their capabilities due to new demand, and rapid prototyping has found extensive applications in the creation of composite materials, moulds, tooling, and plugs to help meet these urgent needs. Additionally, advancements in 3D printing technologies have further expanded the possibilities for prototyping and production.

One sector seeing significant improvements due to these innovations is the marine industry, specifically large-scale rapid prototyping, which has revolutionised the creation of boat plugs, moulds, and components, accelerating processes, reducing costs, and allowing design freedom.

31.4.1 WHAT IS 3D PRINTING MANUFACTURING?

3D printing is a process of creating objects layer by layer using digital models. Unlike traditional CNC machining, where material is removed from a solid block, 3D printing, or additive manufacturing, builds up structures with precision and minimal waste. Large-scale 3D printers extend this capability to produce objects of considerable size, making them ideal for applications like boat building.

31.4.2 APPLICATION OF LARGE-SCALE 3D PRINTING IN BOAT BUILDING

Boat plugs and moulds are essential for the construction of high-quality, customized boats. These components serve as the principal foundation for creating the desired forms and shapes for composite hulls and decks, which are later fabricated using fibreglass or other composite materials. Traditionally, boat plugs are manufactured manually or using CNC machines, both of which can be time-consuming and expensive.

Large-scale 3D printing provides a faster and more efficient alternative, where speed and efficiency are essential. It can produce complex geometries in a fraction of the time required by traditional methods, ensuring rapid prototyping of boat plugs and moulds. Not to mention the reduction of material waste and labour costs, which result in cost-effectiveness, making it a more economical solution, especially for one-off or custom projects.

31.4.3 MATERIALS USED IN LARGE-SCALE 3D PRINTING FOR BOATS

Large-scale 3D printing for boat plugs typically employs materials designed to provide durability and precision. Common materials include

- **Thermoplastics:** Materials like PLA, PETG, and ABS are widely used for their ease of printing and mechanical properties.
- **Fibre-Reinforced Polymers:** Composites infused with carbon or glass fibres provide better strength and rigidity, ideal for marine applications.
- **Resins:** UV-curable resins and epoxies can be used for fine detail and surface finish requirements when needed.

31.4.4 CONTINUOUS FIBRE 3D PRINTING

Continuous fibre 3D printing allows the integration of reinforcing fibres such as carbon or glass into the printing process, creating parts with better mechanical properties. This method is particularly a great solution for producing lightweight and high-strength components in industries like aerospace and sports equipment.

31.4.5 HYBRID ADDITIVE MANUFACTURING

Hybrid systems combine additive manufacturing with traditional machining or other fabrication techniques, enabling higher precision and performance. These systems are ideal for producing high-value composite parts with intricate details.

BIBLIOGRAPHY

'Advanced Composites Perfect Paperback' (Fourth edition). Cindy Foreman. Avotek, 2019.

'Composite Manufacturing Technology (Soviet Advanced Composites Technology Series, 1)' (1994th edition). A. G. Bratukhin and V. S. Bogolyubov. Dordrecht: Springer Science+ Business Media, 1994.

'Composite Materials: Design and Applications'. Daniel Gay, Suong V. Hoa and Stephen W. Tsai. CRC Press LLC, 2003.

'Engineering Mechanics of Composite Materials' (Second edition). Isaac M. Daniel and Ori Ishai. New York: Oxford University Press, 2006.

'Essentials of Advanced Composite Fabrication & Repair' (Second edition). Louis C. Dorworth, Ginger L. Gardiner and G. M. Mellema. Aviation Supplies & Academics, 2019.

'Introduction to Composite Materials Design'. Ever J. Barbero. USA: Department of Mechanical & Aerospace Engineering – West Virginia University/Taylor & Francis, 1998.

'Marine Composites' (Second edition). Eric Green Associates. MD: Eric Green Associates Inc., 1999.

'Resin Transfer Moulding for Aerospace Structures' (1998th edition). T. Kruckenberg and R. Paton. Springer, 2012.

'A Review on the Out-of-Autoclave Process for Composite Manufacturing'. Okunzuwa Austine Ekuase, Nafiza Anjum, Vincent Obiozo Eze and Okenwa I. Okoli. Tallahassee, FL, USA: High-Performance Materials Institute, College of Engineering, Florida Agricultural and Mechanical University (FAMU) – Florida State University (FSU), 2022.

'Structural Composite Materials'. F. C. Campbell. ASM International, 2010.

32 Integrating AI into Composite Materials

Artificial intelligence (AI) makes its great appearance in 2024, and it definitely represents a massive change in several aspects of our lives. Its rapid evolution has marked the rise of a new technological era, having a special influence within industries, since most of them are reshaping themselves to match this evolution as quickly as they can, keeping in mind that AI will not be waiting.

In the world of engineering materials, composite materials are known for their versatility, lightweight nature, and superior strength-to-weight ratios, being at the forefront of innovation in aerospace, automotive, marine, construction, and energy sectors. However, designing, manufacturing, and optimizing these materials is a challenging process that requires a deep understanding of material properties, interactions, and performance under different conditions. As industries push boundaries for lighter, stronger, and more efficient materials, artificial intelligence emerges as a transformative force, which, when applied to composite materials, will extend across design, production, testing, logistics, and management, leading to higher efficiencies and innovations. Here is where the integration of AI into the field of composite materials announces a new era of material science innovation.

32.1 REVOLUTIONIZING OPERATIONS WITH AI

Traditionally, designing with composite materials required a huge amount of engineering and labour hours of experimentation, including tests with trial-and-error approaches. Introducing AI to material design and optimisation means that the use of strong algorithms will be in charge of creating rapid simulations and making predictions or suggestions about modelling to optimise material properties. Thinking of AI as a learning machine, models could be analysed using a vast database to predict outcomes based on fibre orientation, resin composition, and curing processes, with the main goal of significantly reducing development time and cost while providing innovative composite configurations that maximise strength while minimizing weight.

AI systems are also revolutionizing quality control and defect detection, which is an area that could take great advantage of it. These systems employ computer vision and machine learning techniques to detect defects in composite layers, such as delamination or voids, which may be invisible to the human eye. Even using the proper defect detectors and techniques, there is still human interpretation that could lead to an error or a wrong reading. Automated inspection systems will ensure higher reliability, providing real-time adjustments to manufacturing processes and reducing waste.

DOI: 10.1201/9781003565222-32

32.2 AI FOR MANAGEMENT AND SCHEDULING IMPROVEMENT

Optimizing manufacturing schedules is another area where AI could make a significant impact. Intelligent scheduling systems could use AI to balance workloads, allocate resources efficiently where needed, and minimize any potential unpredictable production issues or delays. These systems could base their analysis on historical data and real-time production metrics to ensure efficient operations, even during high-demand scenarios.

When speaking about management, it is also needed to include "logistic and supply chain optimisation" in this conversation, since logistics itself is fully related to estimations, inventory management, and supply chain optimisation. By analysing market trends, production schedules, and transportation constraints, AI could ensure timely delivery of materials and finished products. This could reduce costs and minimise alterations in the supply chain.

Not to mention the structural maintenance of composites; this is another aspect that AI can take care of. Algorithms could be applied to analyse composite structures, capturing sensor data to predict wear and tear or potential failure points. Using AI as a tool to predict maintenance tasks will definitely extend the life of composite materials while reducing inactivity and repair costs.

32.3 USE OF AI TO IMPROVE PROCEDURES

Artificial intelligence (AI) promotes a massive transformation in terms of manufacturing with composite materials by improving their processes, such as precision, as well as applying new concepts of automation.

- Automated Process Control: AI systems could monitor and adjust composite manufacturing processes in real time. For instance, in automated fibre placement (AFP), AI can rapidly adjust parameters such as speed, temperature, and tension to ensure uniformity and accuracy. This level of precision could reduce material waste and enhance production efficiency.
- Smart Tooling and Robotics: Robotic systems equipped with AI capabilities are transforming composite manufacturing. The application of smart tooling integrated with AI can adapt to varying production requirements, allowing custom fabrication with minimal reconfigurations in the design, making the production systems more flexible.

32.4 WHAT TO EXPECT IN THE FUTURE OF AI IN COMPOSITE MATERIALS?

32.4.1 Personal Perspective

From my point of view, the convergence of AI and composite materials announces a new era of innovation. While current advancements and innovations are remarkable, the full potential of AI remains untapped. The next decade will likely see AI systems not only optimizing existing processes but also redefining the fundamentals of composite science and engineering.

For sure, some important challenges will need to be addressed, such as data security, ethical AI deployment, and workforce conversion, but there will still be infinite opportunities and obstacles to overcome until we get a fully production version of it. If we want to dream, we can envision a future where AI-powered composites drive the next generation of space exploration, green energy solutions, and urban infrastructure.

As AI continues to evolve, and as a way of prediction, its role in the composite materials industry will expand in several exciting directions:

- AI Material Discovery: Using advanced machine learning, AI could help find new types of materials with better properties, changing industries like aerospace and renewable energy.

- Sustainable Manufacturing: AI will play a crucial role in promoting the sustainability of composite production by optimizing energy usage, reducing waste, and improving recycling processes.

- Self-Healing Composites: Combining AI with nanotechnology, future composite materials may possess self-healing capabilities, enabled by embedded sensors and actuators that detect and repair micro-damage autonomously.

- Increase of Workforce: Collaboration between human and machine will meet new standards, where AI systems will assist workers in complex manufacturing tasks, improving safety and productivity.

32.4.2 CONCLUSION

In conclusion, AI's integration into composite materials is transforming every aspect of the industry. From improving manufacturing operations and logistics to leading to new discoveries in material science, the synergy between AI and composites promises a future of boundless possibilities.

Index

For Product Safety Concerns and Information please contact our EU
representative GPSR@taylorandfrancis.com
Taylor & Francis Verlag GmbH, Kaufingerstraße 24, 80331 München, Germany

www.ingramcontent.com/pod-product-compliance
Lightning Source LLC
Chambersburg PA
CBHW050524190326
41458CB00005B/1650